THE GARDEN OF VEGAN

HOW PLANTS CAN SAVE THE ANIMALS, THE PLANET AND OUR HEALTH

Why LOVE one
But EAT the other?
LIVE VEGAN!

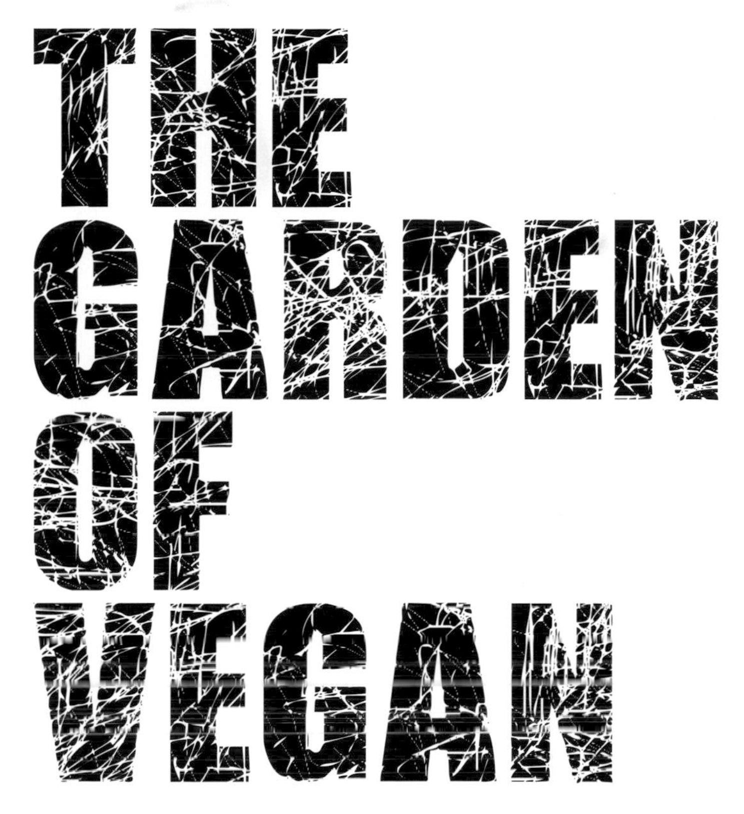

THE GARDEN OF VEGAN

HOW PLANTS CAN SAVE THE ANIMALS, THE PLANET AND OUR HEALTH

CLEVE WEST

This book is dedicated to the 156 billion land animals and up to 2.7 trillion marine animals that are killed each year and the people who speak up for them.*

* See UNFAO, Livestock Primary: Producing animals/slaughtered 2017, www.fao.org/faostat/en/#data/QL (2019), and A. Mood and P. Brooke, 'Estimating the number of fish caught in global fishing each year' (July 2010), www.fishcount.org.uk

Pimpernel Press Limited
www.pimpernelpress.com

The Garden of Vegan

Design by Becky Clarke Design

A catalogue record for this book is available from the British Library.

ISBN 978-1-910258-47-7
Typeset in Avenir and Intro condensed
Printed and bound in China
by C&C Offset Printing Company Limited

9 8 7 6 5 4 3 2 1

Page 2
Cleve at ABP Slaughterhouse, Guildford, Surrey

'MY BODY WILL NOT BE A TOMB FOR OTHER CREATURES.'

LEONARDO DA VINCI

'IF A MAN ASPIRES TOWARDS A RIGHTEOUS LIFE, HIS FIRST ACT OF ABSTINENCE IS FROM INJURY TO ANIMALS.'

ALBERT EINSTEIN

CONTENTS

WHY GO VEGAN?
let me Google that for you...
ANIMAL LIBERATION
ANIMAL

INTRODUCTION

VEGANISM:
'A PHILOSOPHY AND WAY OF LIVING WHICH SEEKS TO EXCLUDE – AS FAR AS IS POSSIBLE AND PRACTICAL – ALL FORMS OF EXPLOITATION OF, AND CRUELTY TO, ANIMALS FOR FOOD, CLOTHING OR ANY OTHER PURPOSE.'

www.vegansociety.com

There was a time when I questioned the importance of my role as a designer. Of course, there were far worse ways of making a living, but I couldn't help wondering if I was contributing anything useful. Two things provided some answers: designing a garden for a hospital and adopting a vegan lifestyle. Exactly how the two experiences are related and what foundation they provide to merit a book may not be immediately clear at this point, but the confluence of these experiences has been pivotal in shaping my view on the role of gardeners in the world and how they might steer it towards a more caring and compassionate one.

A vegetarian for thirty years, my transition to veganism has had a marked effect. I learned more about nutrition than when I studied it as part of a sport science degree. I learned more about propaganda in the food industry and how, contrary to what I'd been led to believe, cows and chickens are far from 'happy'. I learned that animal agriculture is one of the leading causes of climate change and a whole range of environmental catastrophes. I found that disease and illness are more likely to be caused by consumption of animal products than anything hereditary. I learned that a plant-based diet can alleviate some of these illnesses and, in some cases, reverse them. I learned that a drive towards a plant-based diet could offset many of the environmental impacts of animal agriculture and make a positive transition to a more sustainable future. Everything started falling into place. *It's all about plants.* Suddenly, my role as a garden designer didn't seem so trivial after all.

The facts about animal agriculture were so troubling that I briefly considered giving up garden design to become a full-time animal rights advocate. On reflection, I realized I could probably reach more people within the horticultural industry, as I believe that gardeners are well placed to lead the charge with this important movement, which addresses the biggest social injustice of our time.

Animal rights march, 25 August 2018

This book charts my journey from tentative beginnings with my first show garden to an understanding of the restorative power of gardens at a hospital and a slow realization that some of the worst aspects of the Anthroposcene – the current geological age where human activity has been the dominant influence on the environment and climate change – can, if we let it, be mitigated and even fixed by plants. But it's not that easy. Humans don't think logically or compassionately when it comes to their tastebuds. While we're happy taking what gardens can provide to heal us, we're not so good at reciprocating, and the largest garden of all, planet earth, may soon cease to provide if we don't change our ways.

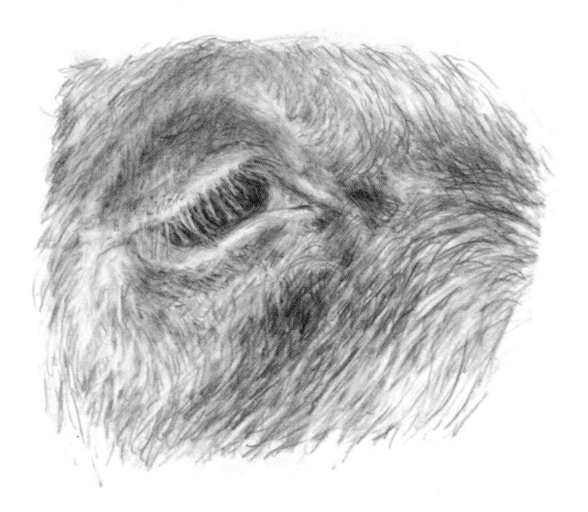

My hope is that people who value gardens and horticulture and benefit from what they provide consider the far-reaching consequences of their actions and food choices, and commit to giving something back to ensure a more sustainable future for all beings that share this Garden of Eden. My hope, also, is that some of my better-known and more influential contemporaries will accept this baton of compassion and run with it, fanfaring the message more eloquently and more forcefully, so that those who think I'm barking up a tree that is too inconvenient for them will sit up and listen.

The scope of the book is, perhaps, a little ambitious. Some chapters warrant a whole book in themselves. Statistics have been taken from sources available at the time of writing, and will almost certainly change by the time this book is published and change again when you find one for sale in your local charity shop. The resources section provides links to websites where statistics are updated from time to time. Readers might dismiss some figures as vegan propaganda and look for loopholes in discussions about the environment, landscape and health to justify the ongoing treatment of animals that is often described as a holocaust.[1] However, one must understand that such propaganda comes from a good place, not from greed, sensory pleasure or profit. It's a desire to live an altruistic life that is considerate and compassionate in the full understanding that our choices have consequences and that these consequences may well involve a victim of violence, torture or exploitation.

While there will be references to things we can do as vegan gardeners, this is not a concise guide to how to garden veganically. There are other books and websites that do this admirably and they are listed in the resources and further reading sections. The book is an attempt to inspire those with a natural predisposition to wildlife and environmentally friendly gardening to go just one step further and make a simple but significant step towards a more compassionate way of life by aligning actions with morals.

Last, if you're uncertain about reading this book because of the

Cleve and Barney

overriding sense of self-righteousness that might come out if it, consider this: most vegans were more self-righteous as non-vegans believing that we had the right to exploit the lives of others for our own enjoyment without any consideration for their feelings, their right to live without suffering and the environmental damage it causes. Vegans are simply accepting that we were wrong to assume this and are now encouraging others to do the same. Adopting a more caring attitude to the world and all living things (plants, animals, insects, invertebrates) can only help improve our relationship with the natural world and, in time, might even have a wider influence on the way we relate to each other as human beings. I live, for the rest of my days, in hope.

1 SEEDS

'BECAUSE WE HAVE VIEWED OTHER ANIMALS THROUGH THE MYOPIC LENS OF OUR SELF-IMPORTANCE, WE HAVE MISPERCEIVED WHO AND WHAT THEY ARE. BECAUSE WE HAVE REPEATED OUR IGNORANCE, ONE TO THE OTHER, WE HAVE MISTAKEN IT FOR KNOWLEDGE.'

TOM REGAN, AUTHOR OF *THE CASE FOR ANIMAL RIGHTS*

Killed at Newman's Abattoir, Farnborough, Hampshire, on 31 August 2016

I wish I could start this chapter by saying how much I loved animals as a child, how I always had an affinity with them and that I was born to become a voice for the voiceless. While I can truthfully say that I did love the companion animals we had during childhood and the farm animals I stroked at Chessington Zoo, I wasn't as kind to insects that crossed my path. It wasn't something I was taught, just an inclination to terrorize anything with six or eight legs. This may have been due, in part, to a traumatic experience disturbing a wasps' nest in my grandparents' garden and being stung repeatedly as I fled when they found their way into my clothes and wellies. In truth, though, it was because I was a nasty little boy. Slugs too were a target for a short while. On catching me stamp on one, my mother's exclamation, 'That's cruel!', was completely misinterpreted as the name for the hapless creature. I remember being confused by her cross face the following day when I proudly told her about several 'Cruels' that I'd managed to dispatch under the heel of my sandals.

My first ten years were spent on Thames Ditton Island close to Hampton Court in Greater London, a romantic place to live where the sense of community within a community was accentuated by the rickety metal bridge we had to cross to and from the village. My brother, André, and I enjoyed the companionship of a dog, a cat, a tortoise and a diverse range of wildfowl that inhabited the waters around us. As soon as we were old enough to behave responsibly on or near water, we were taught to fish. Much of our childhood, therefore, was spent exploring the banks of the River Thames, the River Mole and the ponds of nearby Bushy Park. I loved fishing more than anything else and, while I wasn't particularly good at it, would while away many hours being fascinated by the river, observing nature and the range of wildlife going about their daily routines.

The business of sticking a live maggot or worm on the end of a hook in order to catch a fish did make me wonder about how unpleasant this must be for these poor creatures (not to mention being stuck underwater and then nibbled at), but no one ever suggested that it was unreasonable or wrong. Like most people, I'd learned that it was okay for us to catch fish because people had been doing it for thousands of years. André, more adventurous and always keen to experiment, followed someone's advice to put maggots in his mouth to warm them so they'd wriggle more on his hook. Unfazed, he popped one in his mouth as eagerly as he would a gobstopper and, after a suitable

'REALLY KNOW YOUR SPIRIT AND ESSENCE AND ALIGN WITH IT AND LIVE FROM THAT. IT CAN BE LIKE A WEIGHT LIFTS FROM YOUR SHOULDERS.'

CARMEN O'CONNOR,
The Integrity of Love

length of time, spat it into his palm and pierced the flesh of the writhing maggot with a number 16 hook. Amazingly, on the first cast, he caught a small perch. As memorable as this moment was, I remember being very happy not to catch anything if it meant not having to suck on a live maggot. Another time, while fishing alone, he cast his line so energetically he lost his balance, fell and became trapped headfirst underwater between corrugated sheeting and the landing stage. Only the speedy intervention of a neighbour saved him from drowning.

My brother André and I being introduced to farm animals at Chessington Zoo, Surrey

As was (and still is) customary with coarse fishing in the UK, everything we caught was returned to the water. While most of the dace, gudgeon and roach survived their ordeal, some inevitably suffered from swallowing the hook and our inept, ham-fisted attempts to remove it with a disgorger.[1] It wasn't pretty and I remember being very grateful that fish couldn't scream.

While the latest studies suggest that fish do indeed feel pain,[2] logic makes me wonder why this is even questioned. The ability to feel in some way, however primitive, is the key to survival for all sentient beings. We flee danger because we know it might be painful or at the very least unpleasant. If fish didn't feel, then it's unlikely they could even feed themselves let alone evade predation. But regardless of whether they can feel or not, it turns out that catch and release recreational fishing isn't as harmless as we've been led to believe. Scientists in California found that fish caught using barbless hooks had difficulty feeding compared to those that were caught by nets.[3] Holes in the mouth cause tissue damage which disrupts and weakens the sucking mechanism, not unlike sucking on a straw with a hole in it. In any event, being hauled out of the water with a hook in your face and left on dry land unable to breathe is very difficult to justify if you think about it from the point of view of the victim – the fish.

While my coarse fishing days involved limited harm to the creatures I pulled from the Thames, the trout that I caught as a teenager, in the streams of Exmoor when our family moved to Somerset, weren't so lucky. These were killed with a blow to the head and eaten. Inevitably, during the time spent living in the West Country, conflicting thoughts on animal cruelty began to surface. I was still eating meat in those days, so if an animal was killed quickly for food then it seemed perfectly reasonable, but fox and stag hunting, which was and still is popular on Exmoor (albeit under the deceptive guise of 'trail hunting'), was disturbing. What entertainment could be had from terrorizing a beautiful animal, occasionally chased into the sea before it returned

Going fishing with André and friends at Thames Ditton, Surrey

to shore, only to be savaged by dogs? I was a million miles from being vegan but remember thinking such behaviour was barbaric, cowardly and repugnant. Even if it was cultural tradition, taking part in such an outdated pastime seemed a pathetic excuse to terrorize our native fauna.

Aside from the fish that I killed, my only real hunting experience was for pheasant. One particular pheasant (let's call him Bob) spent most of his day foraging in the coombe that our house overlooked on the edge of the moors. I'm not sure why, but to give Bob a fighting chance I decided to hunt him using a catapult. On seeing him from my bedroom window I'd carefully stalk him to give it my best shot, but he eluded me every time. In the end it was like a very long and boring episode of the animated Roadrunner series without the associated karmic retribution exacted on the ill-fated Wile E. Coyote. The closest I came to killing Bob was when I'd taken almost an hour to creep up on him through bracken and gorse, keeping one eye out for sunbathing adders. When my chance finally came, the leather pouch holding the stone came loose and the catapult misfired. The missile brushed Bob's wing so softly he didn't even bother to fly away. It was the closest I'd ever been to a pheasant, and for the first time I saw just how beautiful it was. Our eyes met. Mine said 'sorry'; Bob's said 'dickhead'. I never went after him again.

I'll end this contrition with two incidents that still haunt me. The first was as a twelve-year-old boy larking about with a friend's air rifle. I took aim at a family of Canada geese and hit a chick in the neck. It wasn't a powerful gun, but the pellet did enough damage to weaken the baby bird and it eventually drowned. I walked home in a vacuum of shame and confusion. I'm sure my friend felt it too but was too polite (or shocked) to say anything. What, exactly, was I trying to prove? Such was the guilt that it's only now, at the time of writing, that I've told this story to my wife, Christine. Needless to say, she wasn't impressed. Twenty years later, fishing in a freshwater dyke with my brother in Florida, I caught and played the most beautiful river bass. It put up an incredible fight (who wouldn't fighting for your life with a hook in your mouth?) and obligingly leapt from the water to let sun glint on its scales to make the experience even more memorable. Once it was landed I couldn't bring myself to kill it and, coward that I am, pleaded with my brother to do it for me.[4] The same void descended. Why didn't I feel like this when ordering fish and chips? What was the difference? We ate the bass that evening out of a sense of duty; I don't recall anyone enjoying it one bit.

The reasoning behind this admission of guilt is not really to get things off my chest (I'm sure that it won't endear me to my vegan friends), but to illustrate how conditioned humans are when it comes to things that are considered 'normal' during childhood. Morality has evolved throughout human history and attitudes have changed as a result, such as attitudes to slavery and women's rights. What might be acceptable at this point in time may one day be completely deplorable.

What with these deliberate acts of violence and all the other animal products I've consumed for over half a century, there's an urgent sense of atonement in my approach to veganism, which, in turn, is the main driver to becoming more active in the fight for animal rights.

DIET

Food has always been a colourful aspect of family life thanks largely to my Anglo-Indian heritage. Weekend visits to my maternal grandmother's house in Harrow during the 1960s were memorable not just for the quantity of food to feed an ever-expanding family, but also for the range of food we were eating. Variety was the spice of life, and there was plenty of spice in that variety. Curries, dals, dosas, bhajis, parathas, puris and pickles. Chapatis, samosas, pakora and pilafs, idlee, kofta and chaat. These savoury delights were finished with desserts of rich and creamy sweet meats, with balls of jalebi, gulab jamun and ruscula swimming in syrup or ras malai (cardamom-flavoured channa) drowning in a sea of cream. Spicy, sweet and infinitely sexier than the bland, overcooked English meals we were served at school or elsewhere, and with the UK only just beginning to learn from Elizabeth David what to do with garlic, our tastebuds were already spoilt for choice.

At school during the mid-1970s, domestic science was a subject which many boys of my age were too self-conscious to sign up for. My early experiences in gran's kitchen, however (and later when my parents bought and managed a small hotel in Somerset), made me oblivious to the general embarrassment and teasing expressed by others. While things have changed immeasurably, there's still a measure of machismo when it comes to food. How many times have I heard men referring to vegans as weak and the alpha-male (who bravely hunted his prey neatly wrapped in the sanitized meat aisles of the local supermarket) take charge of the BBQ saying, 'I could never go without my meat'?[5] While my mates would literally wet themselves at the idea of me as an alpha male, it's very likely that I said exactly the same thing during my carnist years while trying to fulfil a promising sporting career.[6] Mindful that my diet had to provide enough nutrients to allow me to train six days a week, often twice a day, I consumed meat on a regular basis. Considering what I've learned in the last few years, and especially since seeing the film *The Game Changers* (www.gamechangersmovie.com), it's now clear that my non-vegan diet stopped me from fulfilling my true potential.

My dietary habits improved soon after I met Christine and her daughters, Joanna and Stephanie. As a family of animal lovers, they had two cats, Santana and Fritz, and Dicken, a Norwegian Buhund not unlike a small husky. The cats warmed to me; Dicken didn't approve of another male in the household and made his feelings known. I tried not to take it personally, especially as it turned out that (family aside) he didn't really like anyone and had a particular dislike for men on their own, black dogs, cyclists and joggers. A companion animal that doesn't return affection is an odd thing to deal with, but that aside, living with a family more compassionate to animals helped me think about things from a different perspective. Time was occasionally spent rescuing injured animals that crossed our path, including a pigeon nursed at home to recover from a cat attack (not ours on that particular instance) and a concussed fox that had been hit by a car. Seeing it stagger across the road like a drunk, I instinctively picked up the poor fox, wrapped it in my coat and carried it to safety at a friend's house nearby. It was the first time I'd ever been so close to a fox, let alone be able to bury my face into the back of its neck and take a draught of its fur as you would a cat. It's not something I'd recommend,[7] but it was clear from the fox's demeanor that it posed no threat and actually seemed very grateful to be carried to safety.

Another spontaneous act of compassion (and a sign of things to come) occurred while walking with friends across Twickenham Bridge one summer. A commotion in the Thames below turned out to be a pigeon entangled in fishing line floating with the tide in between frantic bouts of flapping. Clearly, it was going to drown so I pleaded with the strongest swimmer in the group to rescue the poor bird. 'Save a flying rat?' he laughed. 'Don't be so ridiculous.' I was left with no option but to go in myself. I'm a poor swimmer and didn't relish making a spectacle of myself; nevertheless I soon found myself stripped to my underpants on the bank of the Thames. The tide was out so it wasn't like I was going to have to swim a long way. Even so I made one last and feeble attempt to get my friend to do the honourable thing, but it was too late – a small crowd had gathered to watch the rescue mission so there was no going back. When the water was waist high, I lowered myself and launched into a gentle breaststroke. Being careful not to cause too much of a wake that would almost certainly finish off the bird, I took my time and swam downstream in an arc so that the bird would float towards me and minimize the distress that I might cause it. As the pigeon came nearer, the full realization of what I was about to do became clear. I would have to hold the bird aloft in one hand and then swim back to shore with the other. The look of fear in the eyes of the pigeon was nothing compared to the panic in mine as I scooped the bird into the air. It could only end badly and, as I took my first gulp of muddy water, I cursed the compassionate instinct that had made me take leave of my senses. Before I could think about treading water (I couldn't even do that very well) my feet touched something solid. I recoiled from it thinking it might be a cable or something equally hazardous that would drown us both, but just as I was about to ditch the bird and save myself I realized it was the riverbed. Both feet were on the floor now and I could stand. I'd swum a few metres out into the middle of the Thames and the water was still only waist high. If I'd thought about it, I could have just kept my knees bent and pretended to swim effortlessly back to the point I'd set off from, but there I was with the water barely covering my crotch. Doing my best to ignore the hysterical laughter and polite applause from the shoreline, I held the pigeon tightly as it struggled to free itself. I don't think anyone knew just how fearful I'd been about drowning, but at least the pigeon was saved and my underpants were intact so, all in all, a relief all round. Ten minutes later, once I'd disentangled a complex mess of nylon fishing line from the bird's wings and feet, it flew free.

These small acts of kindness not only felt good but also broadened my outlook to help me understand that non-companion animals also have feelings, emotions and needs that are worthy of consideration. But that's about as far as it went. Joanna, who had been vegan since the age of thirteen, occasionally spoke to us about the embedded cruelty in the dairy industry, but it fell on deaf ears. Writing about it now I honestly can't explain the cognitive dissonance I had at the time, except to say that years of conditioning by parents, schools and governments led me to believe that dairy was absolutely fine, and so too were eggs. Surely no animals were harmed in the production of these products? Cows had to be milked and it was natural for chickens to lay eggs every day for us. What could possibly be wrong with that?

My decision to become vegetarian in my late twenties came not from a sudden epiphany about animal rights but for health-related reasons. An excruciating experience passing gallstones led a doctor to advise me to cut down my meat consumption. Within weeks of abstaining from meat I noticed an overall improvement in my health, so it made sense to eliminate it from my diet forever. This was welcome news to Joanna, but it was also incredibly frustrating for her that I didn't go the extra mile and give up dairy. Any vegan will tell you that their biggest regret is not doing it sooner and I'm no exception, but my knowledge about animal cruelty and exploitation in those days was non-existent. Even if someone had tried to educate me, it's unlikely I would have paid much attention. From time spent living in rural Somerset and with an aunt living next door to a dairy farm, my opinion was like most others: as long as the animal is raised and killed humanely, I had no problem with it. This disconnect meant that it would take another thirty years to see through the oxymoron that is 'humane slaughter'[8] and accept that the dairy industry is the meat industry and, in many ways, much more exploitative, cruel and twisted as a result.

I did miss a few things initially. Bacon, of course, was top of the list, and giving up my mum's chicken curry was certainly a challenge. As my tastebuds adapted and changed, the cravings quickly disappeared, and over the years the ethical aspects of eating animals started to surface. Veganism was still in its infancy in those days so it didn't even cross my mind to ditch dairy, especially as I had studied physical education and competed at a national level in track and field athletics. During that time I don't recall reading anything about plant-based protein. Had I done more research I'd have learned how a plant-based diet would enhance athletic performance, not hinder it.[9] The Internet and Google were a decade away, so if it wasn't in the college library I wasn't going to make the effort to find it, but American nutritionist Nathan Pritikin had already published *The Pritikin Program for Diet and Exercise* in 1979.

Pritikin's studies found little evidence of heart disease among cultures where a vegetarian diet was the norm. He famously experimented on himself after being diagnosed with heart disease, alternating between the consumption of meat and dairy and plant-based food. He found that his condition worsened when consuming animals and improved when consuming plants, validating his revolutionary theory that diet and exercise (not drugs and surgery) should be the first line of defence against heart disease.

Had I been made aware of Pritikin's studies at the time, I'm not entirely certain that I'd have been open-minded and selfless enough to take on board his findings. Not only did I enjoy consuming dairy products, but also the studies I'd read (almost certainly funded by the meat and dairy industries) supported the general view that a balanced diet should include animals and their secretions. Accepting Pritikin's findings would have turned everything I'd been taught on its head, and I would have had to question the so-called 'hereditary diseases' (heart disease and cancer) that had taken some of our family at a relatively young age. Only now do I look back at our exotic cuisine and see the three main killers in the western world: animal fat, salt and sugar. Lots of it. It turns out that the only thing that was hereditary in our family was diet.

Of course, this is not unusual. Most people since birth have been told that eating animals is completely natural and necessary for good health. Couple that with the fact that it tastes good and the amount of propaganda we're bombarded with every day, and it's easy to see why the carnist badge wearer will find the very notion of veganism absurd.

CULL OR KILL?

Ever since I can remember, even as a pre-vegan, the word 'cull' has annoyed me. Why do we have to sugarcoat the truth? Cull means kill, but because it's not personal and we're just doing it to keep numbers down, we make up a different word to ease our conscience and pacify those who believe that animals have just as much right to live on this planet as we do.

The badger, one of our most beautiful native species, has been a victim of our disconnect with reality. Badger culls between 2013 and 2018, carried out in an attempt to control bovine tuberculosis (TB), killed around 67,000 badgers. It has devastated badger populations, pushing the species to the verge of extinction in some areas, just to placate the dairy industry. The insanity of this was clear to me even in the days when I ate cheese. Bovine TB is about cows (the clue is in the name); the badgers may or may not transmit the disease (the science is sketchy) through going about their natural daily business, which they have done since the Ice Age. What's more, the disease is just as likely to be spread by humans with their outdoor activities enjoying the footpaths that cross farmland throughout the country. Illegal hunting activities are some of the worst perpetrators with horses, dogs, people and vehicles covering more ground than any hiker, yet the badger gets the blame every time.

Now, as a vegan I see the futility even more clearly. For a nation of animal lovers who enjoy the rich diversity of our wildlife, we spend a heck of a lot of time trying to wipe it out.[10] Squirrels are a case in point. Gardeners know they can be a problem, but much of this can be solved by planning and implementing good crop protection. Loved by some, derided by others, the grey squirrel is an interesting adversary for the gardener but wrongly blamed for the demise of the red squirrel, once a British native.[11] In fact, it couldn't be further from the truth. Red squirrel numbers crashed towards the end of the eighteenth century when the pine forests in which they thrive were lost to human activity. With their natural habitat destroyed, their numbers plummeted, and despite introductions from Scandinavia they exist primarily in the north of the British Isles where there are sufficient pine forests to support them.[12]

The grey squirrel was subsequently introduced to Britain from the USA and, being more adaptable, thrived to a point where people think that it is solely responsible for the demise of its red cousin. It also gets blamed for damaging trees throughout the UK. Squirrels are indeed responsible for 5 per cent of forest damage, but according to the Forestry Commission's own statistics, this is an acceptable percentage as the occurrence of bark stripping in the wild is a good thing, providing food and habitat for birds and other creatures.

Sadly, the squirrel is still on the wanted list, especially in the Royal Parks of London where they are just one of many animals killed each year to keep numbers in check.[13] People who come to enjoy these open spaces and the wildlife that live there are generally unaware of the slaughter that takes place each year.

MANAGING NATURE

Evidence-based ecology is beginning to move away from the idea that nature needs to be 'managed' by *homo sapiens*. Animal numbers respond to food availability, so, even in the absence of predators, those numbers are not going to simply spiral out of control. An open and informed debate is needed, ranging from the possible justification for controlling the numbers of animals to the actual methods employed.[14]

Deer, the main feature of Richmond, Bushy and Home Parks, are given supplementary feed to keep them healthy. Around four hundred are killed each year and sold for meat, so in effect they are being farmed.[15] What's more, the numbers of squirrels and crows increase as a result of this extra feeding, and they too are killed as a consequence. Deer populations, especially in parks where they are easily monitored, can be controlled by an immunocontraception vaccine. Porcine Zone Pellucida (PZP), a harmless natural protein, can be used to control fertility in adult female deer and other mammals, and delivered by hand or by darts from a gun. Improvements to the vaccine mean that it can now render the animal sterile for three years, which reduces the amount of darting needed, not to mention the cost. Currently, this protein is derived from freshly slaughtered pigs and is therefore far from perfect from a vegan perspective.

Compared to the amount of wilful, unnatural and often unnecessary damage done by humans to the environment, complaining about what animals do naturally pales to insignificance. We have dominion over animals in terms of intelligence, but we're painfully inept at using that gift to think and live in a sustainable and respectful way to guarantee a future for all life on earth

The human species is without doubt a marvel of evolution, but the simple truth is that we're not as important as we like to think we are and our extinction wouldn't be missed. In fact, as far as the perpetuation of life on earth is concerned, we're not important at all, and every living creature – not to mention the air, waterways and forests – would flourish as a result of our absence. Bees, ants, plankton and other small creatures are, however, crucial to the survival of much of the life on our planet: something our egos would do well to dwell on from time to time.

While our egos would have us at the top of a pyramid of importance in the animal kingdom, an objective view of our standing in the world would place us in a circle with equal status for all beings. In fact, if we are honest about it, we are a nuisance to the planet, and if anyone should be classed as vermin it's us. Unlike bees, ants, birds and bats, we are an invasive species that takes without giving back.[16] We serve no real purpose in that we contribute absolutely nothing to the ongoing survival of other animal and plant species. Unless we see a definite shift towards a vegan world and a broader, more selfless outlook, this misplaced sense of supremacy could very well prove to be the downfall of not just the human race but many other plant and animal species that share the world with us.

EPIPHANY

As a gardener, I found myself drawn to the organic movement, which takes a more holistic approach to gardening, minimizing the impact on wildlife. When taking on an allotment at the turn of the millennium, we decided from the outset to do our best to respect all living things that wanted to share our allotment with us. Apart from one weak moment, nothing was killed deliberately.[17] The jigsaw finally came together when Otto, Christine's grandson, chose to become vegan at the age of twelve. His sister and parents soon joined him, and for the first time we allowed ourselves to give the concept of veganism serious thought. Our dairy consumption wasn't excessive, so it was relatively easy to put an end to that. Until that moment the only thing that had really stopped me (this is so pathetic it's embarrassing now to admit it) was that I enjoyed cow's milk in tea. 'A nice cup of tea and a jolly good sit down' was a habit that I'd repeated at least three times a day for over half a century. The psychological addiction, therefore, was almost as strong as the taste addiction. While Christine had never liked milk, her craving for cheese (despite always feeling uncomfortable after eating it) was the only thing that stopped her from being vegan. Eventually, after running down my supply of milk, I decided to give it a try. My vegan birthday occurred in June 2015, and a week or so later, once the eggs and cheese had run out, Christine joined me.

Despite the fact that Joanna had produced healthy vegan children, we still had some nutritional concerns. What would be our main source of protein if we didn't consume dairy products? What about calcium? Were we going to have to eat lentils for the rest of our lives? As an ex-sportsman who had studied nutrition at college and a gardener who had worked with plants for much of his life, it's odd – and slightly alarming – that I didn't know there was protein in an apple. I didn't know there was protein in oranges, bananas, wheat, rice, potatoes, cabbage, spinach, cauliflower, carrots, swede, kale and broccoli, to name just a few of the 20,000 or so edible plants on this planet. Such is the marketing power of the meat and dairy industries. Like most people I'd been conditioned since birth into thinking that the most vital source of protein comes from meat, dairy, fish and eggs.

Within weeks of giving up dairy, something happened that I hadn't anticipated. I felt healthier. I couldn't put my finger on exactly what it was, but I had noticeably more energy and no post-lunch drowsiness that I'd usually get after a cheese sandwich. It was a genuine surprise and, as time went on, I noticed other changes. Hayfever, which had cursed me since the age of fourteen, hardly affected me the following summer, and an allergy to several kinds of fruit (a kick in the teeth for a vegetarian) disappeared completely. Of course, that could be a coincidental, age-related change, but the minute I gave up dairy products there was noticeably less mucus, which had to be

> **'MORALITY IS DOING WHAT'S RIGHT REGARDLESS OF WHAT YOU'RE TOLD. OBEDIENCE IS DOING WHAT YOU'RE TOLD REGARDLESS OF WHAT IS RIGHT.'**
>
> H. L. MENCKEN

a good thing. In addition, arthritis in a big toe eased enough for me to wear certain kinds of shoes for the first time in years, and as I write these words, I'm toying with the idea of running again.[18]

With the majority of Christine's family now vegan (nine of us around the dinner table at Christmas), I often think of how patient Joanna had been with us and how lame our excuses for not going vegan earlier must have sounded, especially as we can both cook and had previously enjoyed preparing and eating vegan food when Joanna came to eat with us.

After just a month it was clear that we'd made the best decision of our lives. The transition was much easier than expected, the range of food we were now eating made mealtimes more interesting, and we felt altogether healthier by not consuming the secretions of animals. We had finally weaned ourselves off the teat, and even without considering the ethics behind veganism, there was no going back.

ETHICS

On learning more about veganism, things took another turn that I hadn't anticipated. It started when I watched three film documentaries, *Forks Over Knives*, *Cowspiracy* and *Earthlings*. *Forks Over Knives* looks at how food affects our health; *Cowspiracy* highlights the environmental damage caused by animal agriculture; *Earthlings* shows the full spectrum of our dominion and exploitation of animals and pulls no punches. Each film was met with a mixture of emotions ranging from shame and guilt to disgust and horror. I'd been living in a completely different world to the one I knew now, and on seeing graphic footage of what goes on in the dairy and egg industries, any last yearning or desire for milk, cream, cheese, yoghurt and eggs disappeared forever. The neuropathways that associated such food with pleasure had been cauterized and factory settings had been restored. The dairy industry wasn't about pleasure at all; it was about pain, suffering and nothing short of sexual exploitation and oppression.

Understanding the plight of animals by trying to imagine myself in their place and coming to terms with the deceit and lies that I'd been fed my whole life was, as it is for many vegans, a life-changing experience. There was the realization that not only have you unwittingly been contributing to suffering on an unimaginable scale, but also the consequences of your choices as a consumer are contributing to a number of ongoing environmental catastrophes around the globe. If that wasn't enough, you then have to deal with family, friends and colleagues who not only refute, ignore or ridicule the plight of enslaved animals being tortured, raped, experimented on and murdered, but who also seem ambivalent to the negative effects that these actions have on the environment, health and a number of human rights issues. At this point, when your values (based on compassion, kindness and altruism) are being called into question by people you love and trust, the world can suddenly seem confusing and even hostile.

It's fair to say that my foray into vegan ethics caught me off guard. If I'm honest, I'd go even further and say it derailed me. So shocking was the reality of what we do to animals that I found it hard to process. The only thing more difficult than acknowledging one's own cognitive dissonance is understanding the cognitive dissonance of others and especially those closest to you. This occasionally had me hunkering down in a place I didn't know but one that is all too familiar with animal advocates. Author and advocate Kim Stallwood has a name for it: 'Confronted in a world of unremitting violence towards other animals, where others either fail to see what you witness everywhere or don't act on their own conscience, it's easy for advocates to seek refuge in what I call the Misanthropic Bunker. We've all done it: no matter how positive or can do our attitude about convincing others to become vegan or creating social change may be. And here is another truth about animal advocacy: that the retreat may be a necessary way of coping.'[19] After what seemed like a year of soul-searching, I decided that the only healthy way of dealing with it was to try and raise awareness for the victims of this unnecessary system of oppression and how this same system was damaging the environment, something that the horticultural industry is generally interested in protecting.

My first experience of the coming mixture of reactions to this decision came within a week. After much deliberation over whether to get rid of my car because it had leather seats, I settled on the idea that it wouldn't serve any purpose to incur the expense of selling and buying a vehicle, and seeing that I'd only bought it a year or so before, I should keep it and make sure my next purchase was more ethical. Literally minutes after making my mind up on this dilemma, someone stopped to speak to me as I pulled into the drive. 'I hear you're now vegan,' he smiled in a mocking kind of way. 'What about the leather seats in your car?' In other words, 'Don't expect me to give up funding the violence in my bacon sandwich if you're sitting on a dead cow.' Of course, he had a good point, but he could have said, 'I hear you're vegan. Good on you. I guess we should all be scrutinizing our choices as consumers and make more informed, ethical decisions. I find the notion of veganism challenging but I'll look into it.' It was a useful early lesson. I knew from that moment that if I was going to be a voice for animals, the car had to go along with leather shoes, wallet and belt, not to mention woollen or silk items.

It costs money to replace non-vegan posessions, so it's not always possible for vegans to buy a new wardrobe immediately. It seems reasonable therefore to wear things out and buy cruelty-free items when they can be afforded. As an advocate, however, making a stand against animal exploitation while wearing leather shoes or a woolly jumper leaves you vulnerable to ridicule from non-vegans looking for the tiniest loophole to catch you out and use it to justify their choice.

Being active in raising awareness about issues associated with animal agriculture wasn't the path I'd had in mind for my vegan journey, let alone my journey as a gardener, but it was impossible to ignore. Until that moment I'd always considered my role as a garden designer as a slightly frivolous occupation, but combining my interest in organic gardening, wildlife and the wider environment with vegan ethics, I now had something more important than all the accolades I'd received at flower shows. A sense of purpose. It's made me re-evaluate my role and how best to utilize the rest of my days on this planet, which is currently facing an ecological emergency unlike anything that could have predicted just a few decades ago.

While much of this book highlights many environmental and health reasons for choosing a vegan lifestyle, my hope is that it won't detract from the overriding message – that animals have just as much right to live as we do and that they shouldn't be killed or exploited unnecessarily. This is the unshakable foundation upon which the vegan movement has been founded (which frankly should be reason enough), but unless people watch graphic footage of farming practices and slaughter for themselves, it's very difficult communicating the facts about animal agriculture in a meaningful or intelligent way. The environmental and health reasons associated with the farming and consumption of animals, therefore, is a convenient back-up to the moral argument (one that I hope will engage the horticulturally minded reader), but even this isn't easy. Despite irrefutable documentary footage of animal abuse or peer-reviewed evidence about the negative impact of animal agriculture on the environment and our health, just one tiny seed of doubt among a whole forest of truths is all that the profiteering industries need to sow to make people ignore or justify the unimaginable atrocities that animals suffer for a brief and instantly forgettable taste sensation.

EXCUSES

For an intelligent species, *Homo sapiens* can be stubbornly obtuse when it comes to arguing a case for our tastebuds, but perhaps most disturbing is our selfishness and/or inability to empathize with what we know to be a sentient being. The overwhelming body of empirical evidence shows that animals (including the ones that are farmed) feel pain, fear and are self-aware. They have complex social lives and have the ability to communicate in their own unique way. If the opposite were true, that animals don't have feelings, don't feel pain or are incapable of an emotional response, then there wouldn't be any need for welfare laws. Despite the fact that welfare laws are often flouted or ignored, the fact that they exist is proof that the vast majority of people understand that all animals are sentient. The only possible reason for suggesting otherwise is to fuel the notion that animals are unthinking beings whose mere existence serves nothing more than the benefit and pleasure of humans. Animals are even more disadvantaged than the women, children and slaves who were once oppressed and denied their rights. Their physical differences numb our ability to empathize with them, and their inability to articulate their objections puts them at the complete mercy of the oppressor – humans. Animals really are the voiceless victims of this world.

MORALITY

One of the main reasons non-vegans find veganism so objectionable is that their morality is being called into question. If people are inherently good (as I'm often reminded), then being told that you are funding animal abuse on a regular basis is a difficult truth to accept. Few would pay for someone else to stab a dog or a cat to death, so why is paying for someone to stab a cow, a pig or a sheep any different? Of course, no one should be accused of being a bad person if they've been taught from birth about the three Ns of animal consumption: 'normal, natural and necessary'.[20] The aim of most vegan advocates, therefore, is not to shame (after all, most of us were once carnists ourselves) but to educate. Those set in their ways will find this challenging, but younger generations, with the inclination to question things they are taught, tend to be more open to change and a more compassionate world. If a child is taught from an early age to respect all life, there's also a chance that such a compassionate outlook is more likely be extended beyond their own species as an adult.

EGO

Have you noticed that when a shark kills a human it's reported widely as a shark attack, but when a human hunts a shark it's called sport?[21] Furthermore, if a shark or any other animal attacks a human, there's a good chance that it will be hunted, either as revenge for the attack or because of the notion that killing the creature will stop further attacks. When we kill the animals we farm or hunt, we often use the notion of 'the food chain' to justify it and that we're apex predators in 'the circle of life'. This perceived dominion over animals is so engrained in our psyche that attacks on humans by animals are taken very seriously. No mention of the food chain or the circle of life now. This is personal. Such incidents naturally cause much grief and pain for those affected and often make headline news across the world.

So, when I decided to become an animal rights advocate, it was inevitable that I'd annoy a few people along the way. Anyone who's been taught that they have an inherent right to do what they want with someone's life actually thinks that you are trying to take something away from them.[22] This assumption that human life is sacred and beyond reproach when it comes to satisfying a taste sensation is a tough nut to crack, and it's been interesting seeing those with enough compassion in their hearts to give it some thought and those who steadfastly resist the notion of compassion if it means change. It's not unreasonable to suppose, therefore, that those with the biggest egos are the least likely to go vegan. While a degree of selfishness is undoubtedly hardwired into our DNA as a survival gene, it's not a particularly attractive character trait for the twenty-first century, especially when you consider that humans are the most destructive species on the planet and that our extinction would benefit not only the planet itself but everything that lives on it with us. The only effective angle when ego blocks logic is personal health, but only if the ego is ready to accept a notion that someone else might be right.

There are many counterarguments to vegan ethics, but 'morality is subjective' is perhaps one of the most baffling (not to mention alarming) comebacks from non-vegans looking to defend their choice to eat meat. Such a statement means that the notion of right or wrong is irrelevant and that it's simply a matter of personal choice, even where there's a victim involved. If morality is subjective it means that any behaviour is acceptable. For example, it would be okay to inflict violence on someone because it's your choice to do this. It would also be okay for someone to inflict violence on you because that was their choice. Clearly this isn't the case, so when such an argument is used it implies that animals don't think like humans, don't reason like humans, don't have a moral code like humans, making it easier for us to suspend our own moral agency to justify exploiting and harming them. Aside from the fact that animals do think and feel and, in some cases, show signs of making decisions based on the concept of altruism or compassion, should the assumption that they lack these abilities, or at least the ability to express them, give us the right to exploit and harm them for an unnecessary reason?

DESERT ISLANDS

The list of other loopholes and excuses is, as you might imagine, exhausting, from 'where do you get your protein' to 'what would you do if you found yourself on a desert island with only a pig to eat?'[23] While the question about protein is an easy one to answer (see page 109), the desert island scenario is real and one of the most likely misfortunes to befall ethical vegans on a regular basis. Look inside any vegan's backpack and you'll find an inflatable lifejacket, a flare, a whistle and a book on edible tropical plants. I've been marooned five times since going vegan, largely because I never remember to update my satnav. To date, a simple tracking device has saved my bacon (and, more appropriately, the pig's) on each occasion. My only real concern is that rising sea levels caused by climate change (in which animal agriculture plays a leading role) means that there will be a sharp reduction in the number of desert islands available to be marooned on in the first place. It's something I worry about every day.

Now, apart from the bit about me not updating my satnav and rising sea levels, none of what I've just written is true. However, such unlikely scenarios are often used by non vegans to justify their right to eat animals. This is what vegans are up against. Andrew Kirchner answers this with a question of his own:

If you were alone on a deserted island with a pig, would you eat the pig or starve to death?

Hmm. If you were not alone, living on a planet with 7 billion people, had access to unlimited fresh fruits, vegetables, nuts, beans, and other healthy foods, and knew animals suffer and die horrible deaths so you could eat them when you don't need to eat them to survive, would you continue to eat them? The difference between our questions is that your scenario will never happen and mine is the choice you face right now. Which do you believe is worth answering?[24]

PLANTS FEEL PAIN

Another argument, and one that should appeal to gardeners, is that 'plants feel pain'. It's often cited as a justification to kill and eat animals when all other arguments fail. In reality, it's merely a defensive distraction to sidestep questions about morality and trivialize the truth about the exploitation of animals and their suffering.

Aside from the fact that plants don't have a central nervous system and, unlike animals, can't flee from danger, it's logical to assume that plants can't feel pain as other life forms do. The work of FBI agent and polygraph expert Cleve Backster that suggests plants have the capacity to feel and even communicate with their environments is fascinating, but before you use this as your go-to argument to justify killing animals, consider this.[25] Our physiology (blunt canines, lack of claws and long digestive tracts) clearly suggests that we have been designed to eat plants rather than meat.[26] Also, if you take into account the huge amount of soya, grass and grain fed to animals before they are killed to be eaten by humans, it's clear that non-vegans cause more plant deaths overall by processing them through other animals first before eating the animals.[27] In other words, if you seriously want to save plants, eat more plants.

The argument inevitably leads to the number of lives (small mammals, insects and birds) lost by growing plants. Our very existence means that other, often unseen, lives will be adversely affected, and vegans do their best to mitigate these impacts as much as they can. Of course, it's unavoidable so, again, in that sense it's impossible to be 100 per cent pure. However, using accidental deaths as an excuse to deliberately exploit and kill others for food or any other reason fails to grasp the basic premise: a respect for all living things. Also, when you consider that it takes 5–20 kilogrammes of grain to make 1 kilogramme of beef and that up to 91 per cent of rainforest destruction is caused by animal agriculture, it's clear that many more lives are lost in the process of feeding farm animals with plants (before consuming the animals) than growing crops that can be fed directly to humans.[28] A vegan diet, therefore, significantly reduces the number of lives lost in both the animal and vegetable kingdoms.

It's frustrating but understandable that non-vegans should look for loopholes to justify eating animals. It helps people disassociate themselves with immoral actions and suppress the compunction when eating creatures that have suffered unnecessarily. And the idea that you are going to miss a particular type of food is a real one. We've become so conditioned to certain tastes, textures and the events and cultural traditions associated with them that the suggestion of giving them up is often seen as a threat or a violation of one's rights. The reality, of course, is completely opposite to what you think might happen. Since going vegan we have expanded our repertoire of recipes and enjoy our food so much more. Naturally, there was a short period where we had to find our feet, checking ingredients and getting to know decent substitutes for all the things we used to enjoy, but the transition was easier than I ever imagined and nowhere near as expensive as anti-vegan propaganda would have you believe.[29] The only limiting factor is finding the time to experiment with a seemingly endless list of new recipes to try.

PROPAGANDA

On taking my first steps as a vegan advocate using social media to sow seeds of truth and raise awareness about animal rights issues,

'IF SLAUGHTERHOUSES HAD GLASS WALLS, EVERYONE WOULD BE VEGETARIAN.'

PAUL McCARTNEY

I was quickly accused of spreading vegan propaganda. There was no point in denying it. The definition of propaganda is as follows: 'Information, ideas or rumours deliberately spread widely to help or harm a person, group, movement, nation, etc.' Propaganda can be selective and, of course, misleading, but it doesn't necessarily mean false or harmful information.

Posting graphic content of slaughterhouse footage and other forms of animal abuse won't endear you to your social media followers, but it does raise awareness of what actually happens behind closed doors and the innocent victims that suffer as a result. It's only when people can see the facts for themselves that they can make a more informed judgement. While previously my dabbling on social media had been mostly flippant observations on gardens and football and corny jokes about my dad, the Internet age presented the means of spreading information normally hidden from the public, and it seemed rude not to take advantage of it.

Supermarkets and the dairy industry know that if people saw the reality of what happens on farms and in slaughterhouses it would be bad for business, so naturally they need to spend millions of pounds hiding it and reinforcing the notion of happy cows and free-range hens. The psychology behind it is genius. It's propaganda at its most powerful (not to mention insidious) and it works. So powerful is the animal agriculture industry in the USA that their ag-gag laws can put advocates in jail just for taking pictures of a farm or slaughterhouse facility without the owner's consent.

Even as a vegetarian, mindful of the shoddy practices on industrialized farms, I happily bought into the notion of organic and free-range. Once I'd seen undercover video footage from accredited organic dairy farms and free-range chicken farms, I was left with a choice. Keep my head in the sand and carry on as a vegetarian knowing I was paying for people to abuse animals or raise awareness about the lies we are being told by large corporations. When I saw for myself the forced inseminations, calves being taken from their mothers, male calves shot within twenty-four hours or raised for veal, cows chained by their feet in sheds and then strung up to have their throats cut, female chicks having their beaks trimmed, male chicks (useless to the industry) thrown live into macerators or suffocated in plastic bags, only then did I understand that the free-range, Red Tractor, RSPCA Approved labels exist only to ease our conscience and not for the welfare of the animals.

The tsunami of propaganda that you suddenly become aware of as a vegan is quite overwhelming. TV, radio, newspapers, periodicals, shops, billboards, hoardings, flyers, leaflets, Internet banners, sports events, food events: the spew of chopped up body parts dressed up and neatly packaged for happy occasions is endless. I'd been blind to it before because I'd grown up in a system where violence was normalized.

Exposing children to the propaganda of animal agriculture is critical for the meat industry to survive. During the infamous McLibel case in 1997 it transpired that children were exposed to 20,000 adverts for McDonald's per year, the clown (Ronald McDonald) and free toys serving as powerful hooks for children.[30] It's no wonder that people feel so indignant when challenged to consider the ethical or health implications of meat consumption. They feel that not only are their rights being challenged, but also their morals as taught to them by people and institutions they trust.

CROSSROADS

Writing this book has been a great opportunity to raise awareness within the horticultural industry, which, as I've already stated, should be at the vanguard of the vegan movement because of the shared focus: plants. If only it were that easy. In January 2019, Sarah Wilson's 'Roots and All' podcast to celebrate the launch of Matthew Appleby's book, *The Super Organic Gardener: Everything You Need To Know about a Vegan Garden*, predictably attracted a mixed response on the *Gardeners' World* Facebook page. Sadly, comments from (hitherto unknown) gardening trolls became so hateful that the thread had to be taken down.

Armed with knowledge that had eluded me as a vegetarian and fired up enough to speak out against the atrocities that animals have to suffer at the hands of humans, I turned to social media. I tweeted, I blogged, I posted graphic footage of animal slaughter on my Facebook page, I showed pictures of miserable creatures, moments before slaughter, on Instagram. I didn't hold back.

Lately, I've tempered my emotionally charged style of advocacy as I realized that, despite having the world of information at our fingertips, people don't like facing up to uncomfortable truths. I've given several talks about veganism to garden societies, garden centres and trade shows, and, encouragingly, a number of gardeners and designers have been in touch for more information. Some have become vegan, a couple are now advocates, and there is a distinctive groundswell of interest in the vegan movement from those who have made the effort to consider the consequences of their choices as consumers.

Since 2014 the number of vegans in the UK has quadrupled from 150,000 to 600,000, and the changes in the food industry are profound.[31] More and more vegan outlets are springing up each year, and supermarkets are falling over themselves to cater for the expanding market. As people wake up to the realities of animal agriculture and its effect on the environment, their health and the animals themselves, it's a

'WRONG DOES NOT CEASE TO BE WRONG BECAUSE THE MAJORITY SHARE IN IT.'

LEO TOLSTOY

If you think eating meat
is a personal choice,
you are forgetting
someone.
*choose vegan

'TO BE "FOR ANIMALS" IS NOT TO BE "AGAINST HUMANITY". TO REQUIRE OTHERS TO TREAT ANIMALS JUSTLY, AS THEIR RIGHTS REQUIRE, IS NOT TO ASK FOR ANYTHING MORE NOR LESS IN THEIR CASE THAN IN THE CASE OF ANY HUMAN TO WHOM JUST TREATMENT IS DUE. THE ANIMAL RIGHTS MOVEMENT IS A PART OF, NOT OPPOSED TO, THE HUMAN RIGHTS MOVEMENT. ATTEMPTS TO DISMISS IT AS ANTI-HUMAN ARE MERE RHETORIC.'

TOM REGAN, *The Case for Animal Rights*

snowball that can only get bigger. My hope is that more influential gardeners than me will reject the exploitation of animals for good and make choices for a more compassionate and sustainable world.

Apart from feeling and being healthier in not consuming animal products, there's an extra feel-good layer that comes from knowing you're not contributing to the mass slaughter of animals and the knock-on effects this has on the environment. It's a feeling I didn't even know I was missing, but, once felt, was as fulfilling as anything I could hope to strive for. The actor Peter Egan sums it up perfectly:

> I am a deeply committed vegan for ethical reasons. I can't stand the cruelty and I do not think I have the right to satisfy my personal appetite by subjugating or destroying the life of any living species on this planet.
>
> I absolutely love being a vegan. It connects my head and my heart in the most profound and creative way and I know that a plant-based diet gives back to our planet in a way that animal agriculture does not.
>
> I gave up eating animals ten years ago after watching the remarkable film *Earthlings*. I have been a vegan for five years. It's a choice I'm glad I made. If you're thinking about it give it a try. You won't regret it.[32]

Bearing witness at Newman's Abattoir, July 2018

2

'IT'S JUST LIKE MAN'S VANITY AND IMPERTINENCE TO CALL AN ANIMAL DUMB BECAUSE IT IS DUMB TO HIS DULL PERCEPTIONS.'

MARK TWAIN

Killed at Newman's Abattoir on 13 December 2017

In 1990, a few years after starting my own landscape and garden maintenance business, I enrolled on a garden design course tutored by the late John Brookes MBE at the Royal Botanic Gardens, Kew. John was not only one of the most respected garden designers of his time, but also one of the best authors and educators who brought the notion of gardens being an extension of the house into sharp focus. In fact, until I was introduced to his groundbreaking book, *The Room Outside*, I had no idea that there was such a profession as a garden designer, so without him it's unlikely that I would be writing these words. Generous and keen for us to get our money's worth, he pushed us hard, gave us a good understanding of the principles of garden design, and encouraged everyone to find their own path in an industry that was reinvigorated by the notion of lifestyle.

It took me several years to pluck up courage to make a show garden, but in 1994 'Homage to the Green Man' at the Hampton Court Flower Show was dedicated to my great-aunt Ivy, who helped me fund both the course at Kew and part of the show garden itself. The Green Man is an ancient symbol of our reliance upon and unity with nature. His foliate face and mouth disgorging leaves is often seen in church ceilings to symbolize the cycle of growth, death and rebirth and was used to smooth and schmooze during the transition from paganism to Christianity.[1]

As it turned out, the garden was strangely prophetic. The Green Man theme was a simple expression of my interest in organics, the cycle of nature and contemporary garden design. Most of the plants had been taken from our own garden and from clients' gardens where no herbicides or pesticides had been used, so I was able to lay claim to the fact that this was the first organic show garden at an RHS event. Like all show gardens, it was quite an effort and, without a commercial sponsor, a drain on my limited resources. But thanks to my aunt's legacy and help from friends and clients (not to mention some stunning timber furniture from sculptor Johnny Woodford), it scored an RHS Silver-Gilt Medal and the George Cooke Memorial Award for the most innovative garden of the show. It was more than I dared hope for and, with good feedback from the local press, helped me get my feet firmly on the first rung of the ladder in the world of garden design.

The late John Brookes MBE visiting my show garden for M&G at Chelsea Flower Show, 2016

However, on spying the *Gardeners' World* presenter Geoff Hamilton standing in front of the garden, legs slightly apart, arms crossed and head cocked to one side, I realized something was up. I was a big admirer of Geoff, his commitment to organics and his work at Barnsdale. He had galvanized our resolve to garden in a more sustainable and compassionate way, so I was a little in awe of him taking the trouble to talk to me about the garden. His stance, however, bothered me and, sure enough, on introducing myself his eyes narrowed and looked rather stern. 'How on earth can you call this an organic garden?' he asked, pointedly, before I could shake his hand. 'Your furniture's made from tropical hardwood!' Taken aback by this gruff introduction I reassured him that the timber he was referring to was, in fact, English elm, not wood from prime rainforest, reclaimed from a Dutch-elm-diseased tree that had been felled in Brighton. It had then been burnt and wire-brushed by the artist to give the impression of something more exotic. On hearing this Geoff relaxed, smiled and held out his hand: all was well. After showing him around the garden, I left him sitting by himself on the accused bench making notes for a rather flattering review later that evening on *Gardeners' World*.

Next up was a formidable-looking hack leaning heavily on a post at the front of the garden, notebook in hand. It turned out to be Graham Rose, garden correspondent for *The Times*, well known for his political incorrectness and loud opinions. 'I've got to be honest, I'm not very keen on all this new-age organic stuff,' he boomed, 'but . . . I like this garden very much!' He then proceeded to recommend gardens that I should visit and tentatively arranged a trip to Little Sparta created by his friend Ian Hamilton Finlay. For someone just finding their feet in the world of garden design, this was hugely encouraging. Both Geoff and Graham were very supportive during the pivotal year after this first show garden, and I was sad to see both of them leave this world within two years of meeting them.

While much of the influence towards organic gardening came from Geoff Hamilton, Christine was advocating a live-and-let-live approach and was against killing anything that wanted to share the garden with us. By the time we took on an allotment in 2001, we were fully on board with organic principles and, while not self-sufficient or experts in any way, managed to enjoy enough of a harvest to make our allotment experience a fulfilling one, despite the frustrations involved protecting it from creatures that wanted a share of the spoils.

Alarmingly, this wasn't necessarily shared by the majority of gardeners we knew, especially at the allotment where slug pellets were often used like mulch and fallout from over-eager glyphosate users affected neighbouring plots like napalm. This 'nuke-on-sight' approach was considered normal, being endorsed by the RHS and programmes like Radio 4's *Gardeners' Question Time* without any consideration for bees, butterflies, ladybirds and many of the pollinators and predators that would also suffer.[2]

Today, while we rely on the local market and supermarkets to top-up any shortcomings at the plot, the allotment still serves as our private

Verbena bonariensis, one of several plants which we allow to self-seed to attract insects

paradise, our place of refuge where just being there is enough to keep us connected to nature, even when couch grass and bindweed constantly test our sense of humour. One very important lesson we have learnt is that growing food isn't always easy and that, if we are to see a vegan future, there is much work to be done to help farmers prepare for the transition and for children to learn the rudiments of gardening from an early age.

DESIGN

Being vegan has made me more mindful of the potential negative impacts of my work as a garden designer. Like any human intervention on the landscape, there's always the potential for destruction to soil structure and the life that it supports when implementing a new project, but what if the positive effects from making a garden are offset by the process involved in making it? At a Cityscapes salon, in 2019, directors Darryl Moore and Adolfo Harrison led a discussion at their London office about the unseen consequences of the design and build process, and questioned how we ensure that the supply and delivery chains do not have net detrimental effects both socially and ecologically. Can we ethically audit our actions to look more closely at everything we do in order to achieve outcomes that balance the economics of running a business, dealing with the demands of clients, and ensuring sustainable and respectful environmental practices? The ensuing debate showed a willingness to engage with such issues and an enthusiasm for taking more responsibility, but asked questions about how to gauge the knock-on effects of landscape works so that, regardless of scale, there will be a net benefit to the space.

What became clear was that clients aren't necessarily engaged with or interested in wider environmental issues beyond the garden fence, and that it was up to us to educate and coerce them into considering the wider implications of the proposed works. For a start, are the planned changes necessary? If not, then how can we justify disturbing existing biodiversity that may have taken several years or even decades to establish? We each have a personal aesthetic preference, and it's highly unlikely that we will stop going about the things we do to our houses and gardens for fear of killing insects and invertebrates along the way, but if the garden carries some historical importance and an existing timeless quality that might be spoiled by a misplaced outdoor kitchen, that at least should be worthy of preservation.

A garden is a living space. Indeed, it is the ultimate 'living room'. Even small gardens are invaluable as carbon sinks, and the more we engage with them and understand the forces that define them, the more sensitive and receptive we will be to the wider environment and the issues that threaten it.

A designer worth their salt should place sustainability at the heart of their practice and create spaces that minimize unnecessary harm and stand the test of time. The effort and upheaval in making them should be mitigated by the overall positive contribution to the surrounding environment over the lifespan of several generations. It was suggested that we should go one step further than sustainable and make our spaces regenerative by taking steps to reduce disturbance, maintenance, water run-off and water consumption; improve soil conditions, erosion control and the use of native flora; and eliminate the use of chemicals, especially those that kill bees and other pollinating insects. It's something I may not have given much credence to as a non-vegan, but once you start joining the dots you become acutely aware of your daily business and the knock-on effects that these actions cause.

At the Maggie's Centre in Cardiff (see page 86), the small thin patch of ground to the rear of the building was so thick with brambles that we'd wrongly assumed it was a flat site. On arriving to reassess the planting scheme, I learned that the intention was to remove the top 30 centimetres (1 foot) of existing topsoil and replace it with new. Three years ago this would have made perfect sense to me. Now, the thought of it jarred. Aside from the cost of taking away the soil and importing new, there was the issue of compaction and destroying the existing structure and invertebrates inhabiting the ground. We decided to clear by hand only what was alien to the site (hardcore/scalpings fallout from the main site) and to grade the existing soil as required, topping up with imported soil where absolutely necessary. The downside to this is that whatever had been growing on the site over the previous few years (brambles mostly) stood a fairly good chance of quickly re-establishing itself. While brambles aren't easy to control, their spread and habit generally make it difficult for other weeds to compete, so with some selective hand-weeding each year the ivy groundcover was able to establish reasonably quickly.

It's also fairly safe to assume that once a client has decided on having their garden designed and landscaped, someone will implement it. Ideally it will be a designer and/or contractor mindful of potential harm that can be done rather than one who has no concern at all. Whatever the scheme and however sensitive it might be to the existing flora and fauna, it will be impossible to work without causing destruction and unseen deaths, but working with the site, topography, soil and prevailing climate will reduce the amount of potential upheaval and make for a better garden in the long run. While many of my early gardens were designed to stand out and challenge preconceptions as to what a garden should be, I've come to appreciate that the best-designed gardens are often understated and where the hand of the designer is largely invisible. Russell Page, one of the leading garden designers of the twentieth century, strived for an air of inevitability in his design and that's a good mantra to heed. If the design feels at home in its surroundings, it will settle so comfortably that the hand of the designer will virtually disappear.

Now's a good a time as any to own up to a failure in my own backyard, and I hope it will serve as a reminder to anyone looking to improve biodiversity in their garden. Two years ago, our boundary fence at home started collapsing. It had been there for over thirty years and would have rotted a lot sooner if it hadn't been for the ivy and honeysuckle that had smothered it, but it had really seen better days. Timber from the oak fence we'd made at Chelsea (and had stored in a yard) started singing to me and eventually the garden got a facelift. While I was pleased with the overall result, something was wrong. I couldn't put my finger on it until one hot humid evening with the windows wide open I noticed that there weren't any moths in the kitchen. While the number of insects coming in had noticeably diminished over the last few years, this year there was nothing. We had taken away a big part of their natural habitat. Neither were there any bats. No hedge means no insects. No insects means no bats. I tried to make up for it by drilling holes in the fence and the shed to attract solitary bees (and they did come in their droves), but the dent we've made in our local biodiversity is significant, upsetting and hugely embarrassing to admit.

The point is that even where intentions are good there can still be significant displacement, so a good habit to get into is to ask whether the damage caused can be justified. The assumptions we make as gardeners, even when the 'improvement' of biodiversity is uppermost in our minds, invariably comes from an anthropocentric point of view. The sense of benevolence from leaving a small portion of the garden undisturbed is real, but what stops us from letting wilderness have the upper hand? In *A New Garden Ethic* (2017), Benjamin Vogt questions our haughty perception that gardens exist solely to please us from an aesthetic point of view:

> A garden is . . . a negotiation between our arrogance and the perceived arrogance of a wildness we constantly battle. Gardens tend to fight back wilderness and make it legible; they mold nature into something we understand and are comfortable with. But in that victory of conquering nature through gardening, we lose a deeper understanding of wildness and ourselves.

Any of us who has had the privilege of spending time in a natural landscape without any obvious human intervention will know the feeling of wonder and timelessness that comes with it. Just speaking aloud can spoil the moment.

Vogt's 'reconciliation ecology' challenges convention and encourages us to think more deeply and compassionately about our intervention with landscape.

> Gardens have deep meaning when they are created and managed to benefit other species, even other humans. Shifting the perspective beyond our own can feel strange and disruptive. And being composed of other sentient, living organisms with their own distinct life processes, gardens have value and meaning beyond their artistic representation. Why we make them and how we design them reveals the extent of our social responsibility, as well as our awareness of how the world works and what ecology really is.

Gardens of all sizes, both urban and rural, have the ability to reconnect us with the natural world and engage with it in a more altruistic way. Of course, imagination (not to mention some selfless commitment) will be needed for an aesthetic that many of us might find unfamiliar, challenging and perhaps in some cases underwhelming in the first few years, but when the health of our planet and the survival of our and many other species is at stake, knowledge and education are vital if we are to continue to indulge our imagination. If that isn't inspiration enough to seek excellence, integrity and sustainability in our industry, I don't know what is.

STOCK-FREE

Unless you know where your food came from and the processes involved in making it, buying food that is free from animal harm can be almost impossible. Even if you buy a vegan pizza, how can you be sure that the flour used to make it was grown on veganic principles? Organically grown food where no pesticides or herbicides have been applied is a good place to start, but manures derived from animals and slaughterhouse waste products are still acceptable to fertilize crops under existing organic standards. By that reckoning, it's virtually impossible to be 100 per cent vegan in the strict sense of the word. However, using this as ammunition to justify the unnecessary slaughter of animals is a cheap shot that simply says, 'If you can't be 100 per cent vegan, then I've got every right to abuse, kill and consume as many animals as I want.' The point is that we have to start somewhere. No social justice movement has ever been successful overnight. The fact is that the world's food systems are so intertwined and embedded with unseen connections to animal agriculture that it will take time to unpick and reset them.

If you have the good fortune to have an allotment or a patch of ground where you can grow your own food, then you have a place where you have a semblance of control and that counts for a lot. Soon after taking on our plot in Bushy Park, west London, and some years before becoming vegan, we were introduced to Iain Tolhurst, a stock-free organic farmer who sells four hundred boxes a week from an 0.8-hectare (2-acre) walled garden and a further 6.5 hectares (16 acres) of open fields on the Hardwick Estate in Oxfordshire. Known locally as 'Tolly', he co-authored *Growing Green* with Jenny Hall and has become one of the most inspirational experts on stock-free organic farming, advising regularly on his techniques both at home and abroad.[3]

As a vegetarian using external inputs such as fish, blood and bone fertilizer and farmyard manure, I was still curious to learn more about the Stock-free Organic Standard, a method of farming that doesn't rely on animal manures or byproducts to fertilize the soil. It also restricts the use of 'natural' pesticides that are unable to discriminate between 'pests' and beneficial insects. Instead, it uses a combination of green manure (sown as a ley or grassland crop where there is space and as a companion crop between vegetables) and long crop rotations to

Iain Tolhurst teaching veganic techniques at Tolhurst Organic

optimize fertility and reduce the damage caused by pests and disease to a manageable level. What Tolly has done is to remove conjecture from the equation. Over the past thirty years his methods have been tried, tested and honed to dispel the myth that animal manures are essential to maintain soil fertility. This closed system – which includes buying and selling locally to reduce environmental impact – is fast becoming a benchmark for veganic farms across the country and around the world.

Tolly's decision to explore the stock-free method was borne not so much out of compassionate grounds as from economics and the difficulty some organic farmers have in sourcing enough organic manure to fertilize the land. Aside from the difficulty tracking the provenance and ethical status of inputs, if crops could be grown sustainably in a closed system (without having to import animal fertilizers), it could potentially increase profit margins. And with one eye on what the future might hold for the earth's energy resources, a system that reduces its carbon footprint through the amount of energy being consumed and the negative impacts that animal agriculture has globally had to be worth exploring.

The stock-free approach means that farms can become almost completely self-sufficient in terms of fertilizer. By adding 'beetle belts' – overgrown strips of land between crops to provide food and habitat for insects – a natural balance evolves where predatory insects feed on slugs, aphids and other 'pests'. There are, of course, huge differences in terms of scale when comparing a farm to an allotment or garden, where damage caused by pests or disease is felt more acutely, but Tolly's methods are proof that by working with nature, not against it, sustainable farming is possible.

Key to the success of the stock-free system is an understanding of soil fertility and the natural ecosystems that sustain it. Tolly's trials in green manuring and the addition of plant-based composts and chipped wood from branches resulted in an increase in soil microbial activity and a more efficient production system that can be more profitable than conventional farms using synthetic fertilizers. It confirmed that, contrary to popular belief, it was possible to improve soil health and grow organically without using animal manure at all.

Trials at the Hardwick Estate and by the Vegan Organic Network showed that plants themselves are the key to soil health: 'Plants alone are the producers of food energy and of soil humus and all animals, including humans, are net consumers. We may convert and concentrate food energy, and thus fertility, in our bodies and wastes, but we are net destroyers of it.'[4] Of course, that's not to deny the fact that animal waste from slaughterhouses improves fertility and yield. However: 'the fertility does not originate from these residues, but rather from the grass and grains which the animals ate. The animals destroy the greatest part of that food energy by their digestion, metabolism and other life processes. Only a small portion of that energy is preserved within the meat, milk products or manure of

'THOUSANDS OF FARMERS AND GROWERS ACROSS THE GLOBE ARE SHOWCASING HOW VEGANIC OFFERS SOLUTIONS TO MANY PRESSING ISSUES. "VEGANIC" IS A COMBINATION OF THE TWO WORDS "VEGAN" AND "ORGANIC", TO CREATE A NEW CONCEPT FOR FARMING. WORLD AGRICULTURE MUST MOVE TOWARDS "PEOPLE NOURISHED PER HECTARE".'

JENNY HALL

'Growing veganically: Because being vegan isn't just about what we eat', Vegan Organic Network, 2019

that animal.'[5] Another reason for not using animal fertilizers is that manure also carries the risk of pathogens that are harmful to humans, including salmonella, E. coli and listeria. There can also be residues from hormones, antibiotics, pesticides, herbicides, heavy metals, worms and other parasites which can be detrimental to human health.

Having applied the closed system to our allotment, so far there has been no significant increase or decrease in productivity and we are still learning. Now that our reliance on free stable manure (delivered to the allotment on a regular basis) has come to an end, we are having to up our game a little in the composting department. Currently we depend on a combination of homemade compost, green manure and woodchip (more details on pages 52–55). Meanwhile, the hope is that as the vegan movement gathers momentum, more stock-free farms will emerge, and eventually veganic staple crops will become widely available.

PESTS AND VERMIN

I've done my best so far to avoid using the words 'pests' and 'vermin', which are usually reserved for the likes of insects, mice, rats and pigeons. Rather than going through each one here, I'm going to point you to Matthew Appleby's excellent book, *The Super Organic Gardener: Everything You Need To Know about a Vegan Garden* (2018). He covers issues and offers advice for a wide range of animals in the chapters on 'Wildlife' and 'Animals and Gardens'.

Naturally and understandably, humans take competition for food very seriously, especially when they are growing it for a living, and are often inclined to take whatever action is necessary to protect their crops. But can we justify the same level of action against anything that might spoil an intended aesthetic in the ornamental garden? While not so critical in terms of survival (food security), gardeners tend to take it personally if their plans for a garden are usurped in any way, especially when many hours of nurturing were involved. I know this because I've experienced it myself. The recent infestation of box tree moth caterpillars (*Cydalima perspectalis*) in the UK devastated buxus topiary in our own garden and at our allotment within days. No sprays were used as it went against my vegan principles, but clients have been happy to use chemicals to keep the infestations under control.[6]

At our allotment, unable to police it as we would if it were on our doorstep, we have inevitably lost crops to a range of creatures, most notably allium leaf miner, slugs, squirrels and pigeons, and, yes, it does drive you slightly bonkers when you add up all the hours spent tending the crops along the way. Like most gardeners, my relationship with slugs has been a rocky one over the years. While no animals have been deliberately killed since taking tenancy of our allotment, I did succumb, in a weak moment before becoming vegan, to feeding slugs to a friend's chickens when the sheer number of them during a particularly wet May was overwhelming. After a few weeks the guilt was too much, and I had to revert to hand-picking and relocating them to the other side of a convenient stream at the back of our plot.

Anyone who has read my first book, *Our Plot*, will also know that human animals have also stolen our fruit and vegetables. Since becoming vegan it's patently clear that in the great scheme of things, humans are by far the biggest 'pest' on the planet, so aside from making sure you do everything possible to protect crops, a little perspective is called for when dealing with non-human pests. The important thing to remember is that these creatures are only doing what comes naturally to them for survival and not just to annoy you.

BEES

Every once in a while, someone will knock on our door at home to tell us that we have a wasps' nest in our roof.[7] It is, in fact, a bees' nest and it's been there for more than five years. Aside from not being able to open one of the bedroom windows and the odd bee finding its way into the bathroom via some secret loft-space passageway, the bees don't cause us any trouble at all. However, with the nest being quite inaccessible at the top corner of the house, it does mean that we are currently unable to put up scaffolding. A beekeeper friend has confirmed that they are indeed honeybees (*Apis mellifera*) and that there's no way of telling how long they will stay, so our plans for decorating the exterior are currently on hold. If we did want to get them relocated, we'd need to access the loft space by breaking through the ceiling in the bedroom or breaking through the roof. Neither is a particularly attractive (let alone cheap) option, so we're happy to have them as rent-free tenants until they decide to move on.

Having had a little experience of beekeeping in my youth, I've always fancied having a hive at our allotment. As vegans not wanting to exploit them for their honey, our interest would be solely in giving them a place where they could go about their work undisturbed. While the bees in our roof seem to be managing quite well without our help, domesticated, non-native bees are susceptible to diseases which can,

Patrinia scabiosifolia, an umbellifer that attracts bees and other pollinators

in turn, affect the populations of wild bee species. The Varroa mite, an external parasite peculiar to *Apis mellifera* and *Apis cerana*, might eventually weaken or wipe out the colony, but for now the bees that land on the blanket weed in our pond for a drink seem healthy enough.

The notion of giving up honey might seem a bit over the top to most people. Even some vegans question whether it's a little on the extreme side, largely because we don't ever think of things from the bee's point of view. It's worth considering that one bee will make one-twelfth of a teaspoon of honey in its lifetime and, collectively, 22,700 trips (some 55,000 miles) to around two million or so flowers are needed to make one pound of regurgitated nectar.[8] Does stealing the food that keeps them alive and healthy during winter seem fair? Is this the best way of showing our gratitude for their important work, especially when it can leave them vulnerable? Spikenard Farm and Bee Sanctuary co-owners Gunther and Vivian Hauk have found that taking the bees' honey weakens their immune system, and they believe that this is what has caused 30–40 per cent of bee colonies to collapse.[9]

As it turns out, whether or not I end up keeping bees, it's not going to make much of a difference in terms of their survival as the honeybee is not currently threatened with extinction.[10] Our concern should lie with wild, solitary bees and other pollinators that do the lion's share of the work when it comes to pollinating.[11] Our love of honeybees and our fascination with commodifying them have distracted us from the importance of these other insects to a point where they are now in jeopardy. Neonicotinoids, a neuroactive pesticide chemically similar

'MANAGED HONEYBEE COLONIES SUPPLEMENT THE WORK OF NATURAL WILD POLLINATORS, NOT THE OTHER WAY AROUND. IN A STUDY OF 41 DIFFERENT CROP SYSTEMS WORLDWIDE, HONEYBEES ONLY INCREASED YIELD IN 14 PER CENT OF THE CROPS. WHO DID ALL THE POLLINATION? NATIVE BEES AND OTHER INSECTS.'

GWEN PEARSON,
'You're worrying about the wrong bees', *Wired* (29 April 2015), www.wired.com

to nicotine, are thought to be responsible for a serious decline in bee populations. The pesticide is used to coat seeds that, in turn, protect the seedlings. However, once the plants mature, the chemical is still present in the flowers which bees and other pollinators come into contact with. The problem is that while pesticides have been tested and developed not to harm honeybees, they are still harmful to wild bees and other insects, so it's with these wild species that concern should lie.

Honeybees aren't naturally tailored to local ecosystems, so they are less efficient than the native pollinators who do the bulk of the work when it comes to pollination. The fact that many of them are solitary and can't be commodified like honeybees means they are constantly under the radar and subsequently overlooked. Until we modify our perception and understanding of how nature works without our help and recognize the importance of all pollinators, not just the ones that we can exploit, we stand to lose the very insects we rely on for our food.

BACKYARD EGGS

The commodification of bees might seem trivial and something that can be ignored, but making exceptions like these paves the way for excuses and loopholes where other forms of exploitation occur. Backyard eggs are a case in point. A lot of people ask me whether it's okay to eat eggs from chickens kept at home and loved as part of the family. Some keep ex-battery chickens (bought for a pound or two) that get to end their days in a loving environment and relative comfort. Many turn up with barely any feathers, and it's a delight to watch them regain their confidence and characters as the feathers grow back.

My family kept chickens when we lived in Somerset, and as a vegetarian I often fancied keeping them at our allotment for a supply of fresh eggs.[12] What I didn't consider was that, regardless of whether money is exchanged or not, sourcing chickens from hatcheries or battery farms for the purpose of consuming their eggs is still supporting an industry where cruel practices take place and perpetuates the mindset that the animals are there for us to exploit. I didn't consider the fact that they had been selectively bred to produce an obscene number of eggs (200–250 menstruation cycles per year) and that this puts a huge strain on their bodies, often causing osteoporosis and shortening their natural lifespan, many dying of complications caused by repeated egg laying. Enjoying the 'benefits' of this exploitation doesn't sit comfortably any more. If I did keep chickens I'd get them from a sanctuary or rescue centre for the sole purpose of giving them safe refuge and a comfortable life. Using this as an excuse or reward to eat their eggs can lead to other exceptions where the commodification of animals is normalized.

It might seem a shame to waste the eggs, but again this is seeing it from an anthropocentric perspective: that the eggs are ours not theirs or some sort of reward for the favour we're doing in caring for them. This shift in perception of what animals mean to us is at the crux of vegan philosophy. We need to acknowledge that animals are here with us, not for us, and accept that they have as much right as anyone else to enjoy a life without unnecessary suffering or exploitation. It's not something we are always aware of (indeed some people who call themselves vegan don't see any problem with consuming backyard eggs), but the minute we assume that the animal, or whatever it's capable of producing, belongs to us or owes us something in return for the care we give it, we create a grey area where commodification takes place no matter how harmless it appears.

So what to do with the eggs? Many sanctuaries feed them back to the hens. It gives them vital minerals and nutrients that are lost during the egg-laying process. Constantly taking their eggs and triggering their bodies into making an immediate replacement continue to put a strain on their bodies and put them at more risk of an early death. Seeing hens as individuals with their own unique characters rather than what they can provide for us is a good place to start.

SOIL AND FERTILITY

One of the most important elements in any garden is, of course, the soil. As far as veganics are concerned, the aim is to achieve long term sustainable fertility. The old adage – look after the soil and the soil will look after you – is worth remembering, and recognizing the type of soil you're gardening is an important first step.

- **Sandy soil** is quick to warm, quick to drain and relatively easy to cultivate. On compressing the soil in your hand, you can see and feel sand particles, and it will hold its shape briefly before falling apart. Its free-draining nature means that it's not as fertile as heavier soils and, more than any other types of soil being used for crops, will need regular additions of organic matter.
- Cold and wet in winter, prone to cracking during drought and difficult to rehydrate, **clay soils** are generally more difficult to work with but have good fertility. Regular mulching with organic compost will improve drainage through the work of earthworms with minimal need for digging. Infestations of pernicious weeds such as couch grass, ground elder and bindweed will be more difficult to eradicate.
- **Silt** is a mix of sand and fine clay particles and associated with the sediment from rivers and waterways. Like sandy soils they are free draining, but hold more moisture so they are easily compacted. They are more fertile than sandy soil but, like all types, will always benefit from homemade compost.
- A mixture of sand, clay and silt produces the magic that we know as **loam**: easy to work, moisture retentive and with good fertility. Depending on the ratio of the mix, it can range from sandy loam to heavy loam. Organic matter will improve soil at any part of the spectrum.

On the face of it, the light, alluvial soil at our allotment is a godsend. The drawback, however, is that it's not as fertile as heavier soils and dries out all too quickly, so we spend too much time watering in periods of drought. Organic matter added to the soil is the answer. A lot of effort, therefore, goes into making our own compost. While we have the space to devote to a number of compost heaps, finding enough green waste to make it can be a problem. Mulching with homemade compost helps to conserve moisture and improve fertility and structure. For the gardener with the luxury of being able to grow their own food, there are several ways of eliminating a dependence on animal products to maintain optimum soil fertility.

COMPOST

Making compost is an essential part of the veganic garden. It's a satisfying process, the art being in getting a sweet-smelling, friable substrate that you can use both for mulch and as a growing medium. Pretty much anything will compost in time, but the general rule is to avoid twiggy matter that will take much longer to decompose. Avoid also adding plants going to seed and the roots of pernicious weeds. Having said all that, I'm pretty hopeless when it comes to screening out weed roots and seeds. We lost the battle with couch grass and bindweed twenty years ago, so I'm not overly fastidious about keeping their roots out of the heap. In fact, when it comes to using the compost, any weed roots that have somehow managed to survive are easy to spot and remove when loading a barrow. Seeds from plants such as *Nigella damascena*, *Borago officinalis*, *Verbena bonariensis* and *Foeniculum vulgare* are always sprouting up here and there but, being easy to hoe, are never a real nuisance. Their nomadic tendancies add charm to the plot when they accidently set seed near crops such as beans where their flowers help to attract more bees. Parsnip seedlings also pop up in random spots and are often left to grow for food or to add a dash of yellow to the plot in their second year when they flower.

One of the best compost heaps I've ever seen was a circular, beehive-shaped form built using bricks. The structure was approximately 2 metres (7 feet) in diameter at the base and 1.2–1.5 metres (4–5 feet) high. No mortar was used and the bricks were spaced to allow air into the heap. The art is laying the bricks in ever reducing circles; it must be built and topped up with compost in stages so it doesn't collapse. In the middle was a long pole which the gardener used to stir the heap and allow air into the centre. The person who showed me this technique had a large garden and had at least three beehive compost heaps at any given time. When the contents had rotted down sufficiently, bricks could be dismantled in part or completely to leave

Brick compost hive at Christ Church CE Primary School, Battersea, London

a giant Dundee cake-like form that could be sliced into as necessary. While many clients have enthused over the idea of 'compost hives', only one (Christ Church CE Primary School, Battersea) has found the space to build not one but three.

Finding enough green waste to fill your own compost heaps and knowing that it comes from an organic source where no chemicals have been used is the biggest challenge. At our allotment we use waste from the plot itself and supplement it with green waste from our gardening activities at home together with a regular supply of kitchen waste.

The amount of organic matter you need to keep the soil in good health each year is probably less than you think. We are told to mulch with 5 centimetres (2 inches) or more to suppress weeds, and this is fine if you have access to this volume of compost. If you haven't you might take comfort from the amount of compost recommended by Tolly and his colleagues at Vegan Organic Network: one barrow load per 10 square metres (108 square feet) applied in early spring.

GREEN MANURE

Green manures (also known as cover crops) are plants grown specifically by organic vegetable growers to fertilize the soil, improve its structure and increase its organic matter content. They can recycle nitrogen, calcium, magnesium, potassium and phosphorus. Plants from the legume family can absorb nitrogen from the air and make it available to subsequent crops once it decomposes. Those left to overwinter help prevent nutrients from being washed away, especially on alluvial soils where this can be a problem.

Grown successfully, green manure can significantly reduce the amount of nitrogen lost over winter (on bare soil) and add the equivalent of 2.5 kilogrammes per square metre (5.5 pounds per 11 square feet) of farmyard manure. Timing is variable where sowing and harvesting are concerned, but generally green manures are incorporated into the ground in spring. Plants are chopped or shredded at ground level before they run to seed and left to wilt before incorporating into the soil, where worms and microbes can work their magic. This can be done using the 'turfing' method where the green manure is undercut to a minimum depth of 10 centimetres (4 inches) and turned over so that no greenery is visible. Decomposition rates vary according to soil quality, moisture content and the weather, but as a general rule they need at least a couple of weeks to decompose before a seedbed is created. The carbon:nitrogen (C:N) ratio of the green manure will also affect decomposition. Carbon-rich Hungarian Rye, for example, will take longer to rot down than clover and will use up more nitrogen in the process.

While growing green manures can mean sacrificing space for vegetables, it is possible to sow them underneath some full-grown crops where the foliage isn't too dense. This includes:

- brassicas + white clover
- cucumber + yellow trefoil, Kent wild white clover
- leeks + cereals
- melons + yellow trefoil, Kent wild white clover
- runner beans + red clover
- squash + red clover
- sweetcorn + red clover
- tomatoes + yellow trefoil, Kent wild white clover.

While it's best to chop down green manures before they mature (especially carbon-rich cereals), plants like phacelia and red clover have such delightful flowers that it's almost impossible to resist letting them have their day and provide extra nectar for visiting pollinators.

WOODCHIP

The use of woodchip as a peat-free propagation material and soil improver is increasingly gaining traction among growers, and with the Department for Environment, Food and Rural Affairs (DEFRA) aiming to phase out the use of peat in the horticultural industry by 2030, woodchip may prove to be both a suitable and a renewable substitute. Peat is non-renewable substrate made from partially decayed vegetation and organic matter that has accumulated over millennia. It acts as an effective carbon sink, but once exposed to the atmosphere, it releases carbon dioxide and methane at an accelerated rate, giving it a high carbon footprint.

I'd already discovered by accident the potential of woodchip as a soil ameliorant after using it to mulch paths between beds at our allotment. Several years of mulching had produced a good depth of well-rotted humus that was added to raised beds after a concern that the woodchip made a perfect habitat for slugs. However, after attending a seminar at Tolhurst Organic and learning about the use of woodchip, it's clear that the benefits of this resource outweigh any of the problems with crop protection. It now figures as a useful asset in our quest to grow food using a stock-free system.

Woodchip ranges from trunk and bark woodchip to chipped-branch wood (CBW), also known as ramial woodchip. Ramial woodchip, which comes from deciduous branch wood in full leaf and with a diameter less than 7 centimetres (3 inches), has a much lower C:N ratio than trunk wood, which, at 300:1–600:1, takes a lot longer to rot down. Ramial woodchip also contains amino acids, cellulose, pectin and starches that benefit the growing of annual crops.

Tolly experimented with windrows of woodchip 3 metres (10 feet) wide, 2 metres (7 feet) high and as long as he had room for. They were turned once after a couple of months and two to three more times over the course of the year. If you have a digger at your disposal, more frequent turnings can speed up the process significantly, but where space is limited, smaller, more manageable piles of woodchip can be left for a year or two. If there's no room for a compost heap at all in

Windrows of ramial woodchip at Tolhurst Organic

your garden, my serendipitous method with deep paths of woodchip can be made and forgotten about until it's ready. Obviously, this is a labour intensive process (especially if you don't have access to heavy machinery), so Tolly has experimented applying fresh CBW (1–2 kilogrammes per square metre/2.2–4.4 pounds per 11 square feet) and raking it into the soil just before sowing green manure (or left bare for weeds to grow through) to decompose over the course of the winter.

Bear in mind that sawdust, wood shavings or tree bark should not be used, as they will lock up nitrogen and affect annual crops for some time. It's also worth noting that local authority green waste and imported woodchip could be contaminated with pesticides, herbicides and other undesirables, so it's worth talking to your supplier first to learn more about provenance.

HUMANURE

Aside from peeing on the compost heap and using our urine as a liquid feed on our tomatoes from time to time, I've not really explored the potential of using human waste as a fertilizer.[13] But having read that 'an adult's urine contains enough nutrients to fertilize 50–100 per cent of the crops needed to feed one adult', it makes sense to consider using this free and plentiful resource.[14] While fresh urine is generally sterile, there is the potential for faecal contamination, so it's best used on fruit-bearing crops rather than as a foliar feed or on root vegetables and low-growing leafy crops.

I've been reluctant to use 'night soil' at our allotment, largely because it involves a bit of organization and diligence with the whole process if it is to be performed safely. 'Night soil' is a euphemism for human faeces, once collected (at night, which gave it its name) by 'gong farmers' during the sixteenth century, and later 'honeypot men' in the nineteenth century. It's been used most regularly by Chinese and Indian cultures in the past and was also used in the UK as recently as the Second World War when fertilizers for home-grown food were scarce. Today, as we question our environmental record and what we can do to improve it, the use of night soil in our gardens remains, as far as I can make out, an untapped resource largely because we are too squeamish.

Pathogens in improperly or incompletely composted human faeces can be extremely hazardous; so too can chemicals that find their way into our sewerage systems. Home waste recycling can reduce the hazards so long as the right processes are adopted and the sewage is given adequate time to decompose completely (about a year in the UK). Again, as a safeguard, fertilizer made from home waste should only be used on fruit crops and not salads or root vegetables where there is the risk of direct contact.[15]

The all-singing and dancing self-contained composting toilets have a large compartment below the toilet with a sophisticated aeration system.[16] Other, smaller systems as advocated by Joseph Jenkins in *The Humanure Handbook* serve only to prepare the waste for a secondary composting in a regular compost heap.[17] Jenkins, who coined the neologism 'humanure', advocates using all household and sanitary waste, believing that rather than being a risk it can actually improve public health and safety. Not only will you reap the benefits in the garden, you will also be saving around 280,000 litres (75,000 gallons) of water a year by not having to flush.

LIQUID FERTILIZERS

Liquid fertilizers aren't allowed under stock-free organic standards, being a short-term quick-fix that contributes no long-term benefits (in terms of organic matter) to enhance soil structure. For the vegan gardener, however, they can be a useful backup, especially for container planting. They can be made by soaking comfrey or nettle leaves in a bucket or rainwater butt for a week or two and using the resulting dark liquid (diluted with ten parts water) on your crops. If you use comfrey, be warned that it will stink like a whale with halitosis so try not to get any on your hands or clothes. Compost tea (I've heard some refer to this as jollop) performs even better as a liquid fertilizer as it contains microbes that will benefit the soil. A porous bag filled with home-made compost is soaked in a container for a few days, strained, stirred to aerate it and then diluted (1:10) before watering plants. Waste solids from comfrey, nettles or compost can be added to the compost heap or directly to the soil.

NO-DIG

The no-dig school championed by the likes of Organic Growers of Durham and Charles Dowding is one I've been wanting to join for years. Letting earthworms do the work of feeding and aerating the soil rather than accidently skewering them and chopping them in half makes complete sense. The reason I'm not completely on that page is because I didn't prepare the ground well enough when I first took on the allotment. Every single scrap of bindweed and couch grass needs to be removed, and to do this on two full-size plots is quite an undertaking without the use of herbicides. However, I was determined not to use chemicals and did the initial clearance by hand. Inevitably, it wasn't thorough enough and I've paid the price ever since.

However, on learning more about no-dig techniques from Stephanie Hafferty (who, with Charles Dowding, co-authored *No Dig Organic Home and Garden*), we've started covering the most infested parts of the plot with cardboard which is then mulched with compost. Any couch grass that finds its way through these layers is then dug up with a hand trowel. In theory, provided we are diligent with hand-weeding, the couch grass will be significantly weakened over the course of eighteen months.

Digging green manure back into the soil is obviously counterproductive in a no-dig system, and it's something we're keen to avoid because where the soil has been undisturbed there's been a noticeable increase in worm activity. Lately we've been leaving chopped-up vegetation on the soil surface to rot down over several weeks before planting.

Planting and harvesting potatoes need special treatment with no-dig gardening. Charles Dowding suggests laying a thicker than usual mulch of compost over your potato bed at the start of winter. Come spring you can then use a trowel or your hand to make a hole for the potatoes without having to dig into the soil below. The idea is that the potatoes will grow into the mulch layer (some earthing up with fresh compost might be necessary) and can be harvested without having to dig over the bed.

'NATURAL AGRICULTURE'

I've always liked the idea of rotating crops. Different groups of plants have different nutritional needs, so a cycle of growing that allows a current crop to benefit from (or tolerate) what's been left by the preceding year's crop makes a lot of sense. Rotations can vary from four to seven years and can be tailored according to the sort of vegetables you are growing. Observing tenants at our allotment who successfully grow the same crops in the same place year after year, however, got me curious as to how they manage to avoid all the pests and diseases that are purported to be more prevalent if you don't rotate. It led me to look into a philosophy known as 'Natural Agriculture', a way of life based on a deep respect for nature encouraging us to live in harmony with its natural life forces.

My interest in Natural Agriculture is a recent one largely influenced by Masanobu Fukuoka's inspiring book, *One Straw Revolution*, and Shumei Natural Agriculture in Yatesbury, Wiltshire. Its appeal lies in its description as a way of rebuilding our relationship with nature without the need for chemicals, hybrid seeds or manures, and that understanding nature's energies can reduce the need for crop rotation and even weeding. Soil analysis at the Shumei farm in Yatesbury has shown an exponential increase in microbial activity and soil health where the same crops are grown in the same place each year using saved seed. The suggestion is that microbes develop a more robust symbiosis with the plant if the same crops are grown rather than having to adapt to a different plant (with different needs) each year. Based on a foundation of cooperation and gratitude, it's a philosophy and an approach to life that rejects profit-driven modern agriculture with its dependence on chemicals and embraces a deeper appreciation and awareness of the interconnectedness of all living things. Such a food system breeds more a sense of partnership than domination and a more respectful relationship with the natural world to reinvigorate traditional farming cultures and environmentally sustainable methods and techniques so that they become second nature, a way of life.

The vision of Mokichi Okada, founder of Natural Agriculture, is world peace. Exactly how that becomes achievable through farming in a particular way may not be apparent, but a child brought up in a culture of truth, cooperation and respect for all living things is a solid building platform from which a more compassionate and respectful world can be built. It may seem like an impossible dream, but the surge of interest in veganism means that people are making the connection and realizing that we need to work with nature rather than against it for a sustainable future.

GREEN ROOF

The sheds I've built at home and at the allotment all have green roofs. That doesn't make me an authority on them, but it always seems rude not to raise the footprint of the shed so that planting opportunities aren't wasted. Early sheds with a shallow substrate (known as an extensive roof system) were planted with *Sedum acre*, sempervivum, and a few plugs of blue love grass (*Eragrostis chloromelas*), which come and go depending on what has managed to self-seed, but overall the roofs were bland and limited in scope. The most recent shed at home is more adventurous with a deeper substrate (known as an intensive roof system) planted with *Sporobolus heterolepsis*, *Dianthus carthusianorum*, *Erigeron karvinskianus* and seasonal bulbs. It's more interesting despite being a little contrived. On our sheds at the allotment (four to date) ivy has been given a platform. These roof spaces, protected with butyl rubber or sheets of corrugated iron, have been loaded with twiggy garden waste to provide a habitat for insects and birds and a loose framework for ivy to eventually find its way into. From an aesthetic point of view they've taken a bit longer to

Ivy smothering garden waste piled on the roof of a shed at our allotment provides important habitat for insects and birds.

get established, but they are cheap to make and have been incredibly successful, requiring little or no maintenance in the first few years while the ivy grows up the wall(s) of the shed. At roof height the ivy's foliage changes from the distinctive five-lobed leaves to its arboreal, heart-shaped form, and it produces yellow-green flowers followed by black fruit. Depending on how much cover you want and how thick the roof canopy needs to be, the ivy may need pruning once every two or three years. The best time for this is late winter before birds start nesting. It will look a bit bare for a while, but it will be in full leaf come midsummer. A mid- to late summer prune is also possible but this will remove flowers (a good source of nectar for insects) and berries (food for birds). The shed needs to be reasonably sturdy as ivy will work its way inside and force timber apart. If this happens regular pruning of the anaemic shoots is recommended. Roof joists also need to be strong enough to support the weight of whatever system you decide on (extensive or intensive). I've always been a big fan of ivy as it's such a versatile plant that will put up with dry shade, but it is surprisingly fussy as ground cover and will die back if repeatedly walked on.

3 HEALING GARDENS

‘THE GLORY OF GARDENING: HANDS IN THE DIRT, HEAD IN THE SUN, HEART WITH NATURE. TO NURTURE A GARDEN IS TO FEED NOT JUST THE BODY, BUT THE SOUL.’

ALFRED AUSTIN

Killed at Kedassia Poultry, London, 25 August 2018

The seed for this book germinated from a talk to raise awareness for Horatio's Garden (see page 79) at the spinal unit of Salisbury District Hospital. Entitled 'Healing Gardens', the talk focused on the story behind the garden and what we learned while creating this space for people with life-changing injuries.[1]

Over time, the talk evolved to include examples of other gardens I'd made or experienced where I'd become more aware of their restorative and therapeutic value. Based solely around my own perception and observation, the talk proved popular with garden clubs and societies, and soon I found myself doing more than I'd bargained for (anyone who knows me will tell you that public speaking gives me the heebie-geebies at the best of times).

Where budgets are tight, I negotiate a deal in return for speaking for ten minutes or so about veganism at the end. This makes it worthwhile; after all, a little stage fright is nothing compared to what animals have the face at the slaughterhouse. Some accept and are genuinely interested; others prefer to pay more and not hear about such things at a garden-related talk. Gradually, though, it dawned on me that what I had to say about veganism had more than a little relevance to gardening. I found that most if not all members of garden societies believe themselves to be passionate environmentalists at heart, so it made sense to tailor my vegan message to the catastrophic effect of animal agriculture on land, water and climate change.

For a while it was hard to gauge the success of this approach (I'm always grateful to the handful of people who offer words of encouragement after the talk while others leave more quickly than a vegan walking through the meat or cheese aisle in the local supermarket), but lately there's been a shift to a more positive response from the gardening fraternity. In September 2018 the Plant Centre at Hortus Loci hosted the UK's first ever Vegan Garden Festival with talks, food and a vegan outreach team offering guidance and information for anyone interested in vegan issues. Even the Royal Horticultural Society has indicated a tentative acceptance that times are changing by allowing plant-based food outlets at its shows and events. With the shared USP – plants – it makes perfect sense.

Gardeners are well placed to make the connection with veganism because we've been reaching out to plants for as long as we can remember and depend on them for our survival. Many people, gardeners or not, intuitively seek the comfort of nature at times of stress. Is this because of the benefits of exercise, aromatherapy and the overwhelming assault on the senses that serve as a useful and pleasant distraction? Or is it because the very notion of nature – the cycle of life, death and rebirth – is so engrained in our psyche that, whether we are conscious of it or not, we reach out to this constant as a comfort blanket

'I LIKE GARDENING – IT'S A PLACE WHERE I FIND MYSELF WHEN I NEED TO LOSE MYSELF.'

ALICE SEBOLD

Bupa garden, Chelsea Flower Show, 2008

when we need it most? Either way, while there may not be enough documented evidence about the restorative power of gardens, it is slowly being accumulated.

This notion of the garden as a place of refuge is nothing new. Ever since the first Paradise gardens of Persia provided sanctuary from the harsh realities of the desert, we've used gardens for recreation and to excite, refresh and restore the senses. Ambitions may have changed over the years in terms of scale, but the key ingredients of a garden remain the same: shelter, shade, water, plants, scent and food. In the UK, gardens were an important feature in psychiatric hospitals of the Victorian era, a trend that succumbed to cuts in space and budgets, not to mention the fixation with cleanliness and sterility. More recently, the therapeutic value of gardens has seen something of a renaissance as medical experts understand the potential restorative effects of outdoor spaces and the long-term benefits for patients in terms of faster recovery times and a reduction in prescribed medication.

In a National Gardens Scheme questionnaire, 39 per cent said that being in a garden makes them feel healthier; 79 per cent believe that access to a garden is essential for quality of life. And in an age of information technology overload, gardens are becoming increasingly important as places to reconnect with nature. For the visitor, they can be otherworldly places in which you can lose yourself; for the gardener they are places where the rhythms of digging, pruning, raking and

'HERE ARE SOME OF THE ESSENTIAL TAKE-HOMES: WE ALL NEED NEARBY NATURE; WE BENEFIT COGNITIVELY AND PSYCHOLOGICALLY FROM HAVING TREES, BODIES OF WATER, AND GREEN SPACES JUST TO LOOK AT; WE SHOULD BE SMARTER ABOUT LANDSCAPING OUR SCHOOLS, HOSPITALS, WORKPLACES AND NEIGHBORHOODS SO EVERYONE GAINS. WE NEED QUICK INCURSIONS TO NATURAL AREAS THAT ENGAGE OUR SENSES. EVERYONE NEEDS ACCESS TO CLEAN, QUIET AND SAFE NATURAL REFUGES IN A CITY. SHORT EXPOSURES TO NATURE CAN MAKE US LESS AGGRESSIVE, MORE CREATIVE, MORE CIVIC MINDED AND HEALTHIER OVERALL.'

FLORENCE WILLIAMS,
The Nature Fix: Why Nature Makes Us Happier, Healthier, and More Creative (2017)

weeding can create an almost trance like state of working meditation. We can spend a day working in the garden and come away exhausted but with an amazing sense of well-being.

If I'm feeling particularly irritable or grumpy (which is most of the time according to Christine), a trip to the allotment or a walk in the park will usually fix it. Of course, any physical activity can have the same effect, but there's something about the connection with nature – the alchemy of things that grow and the sense of achievement in nurturing something from seed to plate – that elevates it to something more meaningful, even when things don't go entirely to plan.

Sue Stuart-Smith is a psychiatrist and psychotherapist who finds that gardening helps her unwind after a hard week in the consulting room. In her book, *The Well Gardened Mind* (2020), she observes that 'renewal takes place so naturally in the plant world, but psychological repair does not come so easily to us.' She highlights how 'in the secular and consumerist worlds that many of us inhabit, we have lost touch with the traditional rituals and rites of passage that can help us navigate our way through life's uncertainties. Gardening itself can be a form of ritual and through an active participation in the growing cycle of life, the mind can internalise something of nature's capacity for renewal and regeneration.' She goes on to explain the importance of the tranquillity we find in gardens: 'When we enter a garden, we can escape from the noise within our own minds and the background noise slips away too. Plants are less frightening and challenging than people, they don't have thoughts or make judgments about us. We are free from the complexity of human relationships, so that a garden can be a more accessible way of reconnecting with our life giving impulses.' Working with plants and attaching to place can be a source of stability; in particular, she believes, it is the 'magic' of making things grow that fosters a sense of self-worth. 'I like to think this green-fingers illusion can act as a psychological growth factor and help counteract a sense of impotence that can overcome all of us at times but especially after a trauma or mental breakdown.'

Twice in the last ten years or so the garden has come to my rescue. The first was when an unknown tropical infection had me confined to bed for a month or more. On the occasional foray into the garden there was no magical cure but a welcome calming effect, one of acceptance of a process I had no control over. The second, while grieving the loss of a parent, was a combination of distraction and memories of shared experiences within the garden context. Tasks in the garden could only be ignored for so long, and some of the most mundane were actually very helpful as a gentle reminder that life goes on. Only a garden can speak to you in this way without it sounding cold and insensitive. Gardening in exactly the same space that had been shared just weeks earlier also accentuated a notion of their presence and, in turn, a sense of continuity. While raw in terms of being able to process, it was nevertheless a handle on which to tentatively hold during fragile months ahead.

My first conscious experience of the therapeutic value of gardens happened by accident. I worked for an art dealer in west London and would visit an aged aunt during lunch breaks to help with the odd chore, often in her garden, which had been somewhat neglected since the death of her husband. Something clicked and I found myself enjoying the company of plants and wildlife more than the office. Soon, thoughts of an outdoor life were too strong to resist, and before long I'd given up my job and started a garden maintenance business, working for my aunt one day a week in return for using some of her tools.

A trip in the wheelchair after lunch for fresh air, weather, scent and sound saw the stress and strain of old age fall from her shoulders as soon as we entered the garden or nearby Chiswick Park. Young and impatient to get back to work, it was only later that I fully appreciated how much these brief moments must have meant to her. Even on the coldest, bleakest of winter days when all I could do was show her a miracle of a rosebud that had somehow prospered against a sheltered, south-facing wall, it was enough to light up her day. Her garden was her universe, not just for the physical, sensory elixir, but also for a lifetime of memories associated with it.

While the experience of the power of gardens is varied and personal, the following examples might offer some insight into what we take from them and how they might be perceived as 'therapeutic' or 'healing'.

Horatio's Garden, Salisbury Hospital, 2012

ALLOTMENT

We were lucky enough to have a small town garden, but had always yearned for a larger space in which to flex our muscles. Taking on an allotment at the turn of the millennium proved useful in satiating this desire and, of course, allowed us to grow our own food. Our timing was fortunate. Within five years of taking on two full-size plots, the trend for Grow Your Own was in full swing. Soon there were waiting lists for allotments throughout most urban areas of the UK and those earmarked for development often made the national press.

Allotments are the complete antithesis to show gardens. For a start, they are real and have a shelf life of more than a week. They don't get the same level of scrutiny and they are much cheaper to make. They are, however, for the long term and the amount of time invested in one involves real commitment.

A ten-minute drive, fifteen-minute cycle or thirty-five-minute walk, the plot is just a little too far to police as we'd like (from both humans and animals). However, the extra effort involved to get there elevates the experience to something akin to a trip to the country or a weekend away, so, despite the frustrations associated with allotments, the restorative effects of this connection with nature, biodiversity and community shouldn't be underestimated.

The allotment also serves as a gym. While not exactly yoga or Pilates, the physical aspects of gardening have been a real bonus. As long as measures are taken to lift, dig and work in a balanced and methodical way, it can be hugely beneficial to one's strength and cardiovascular health. The only danger is overdoing things when you haven't acquired a certain level of fitness or getting distracted by the infinite number of tasks that need doing and forgetting to eat and stay hydrated. Studies have also revealed the presence of anti-depressant microbes in the soil. These cause cytokine levels to rise, resulting in the production of higher levels of serotonin. The bacterium is said to increase cognitive ability, lower stress and improve concentration. Gardeners inhaling the bacteria can enjoy the natural benefits for up to three weeks.[2]

Our plot, which had been left untended for some years before we took over, has been organic from the start of our tenancy. We also adopted a no-kill policy.[3] This was some years before our vegan pledge, but it sat well with us that creatures were not subjected to the sort of chemical annihilation that we witnessed elsewhere at the allotment. Our interaction with wildlife at the plot continues to give us as much pleasure as growing food, and in the way of making the connection to the natural world, it's been a wonderful resource for visiting children.

Summer harvest
from our plot

ABOVE A few of the 30 or so espalier fruit trees at our plot. RIGHT Deer feature in the adjacent Bushy Park. Unfortunately they are killed to keep numbers down and to make money from their meat.

'WE DON'T EXPERIENCE NATURAL ENVIRONMENTS ENOUGH TO REALIZE HOW RESTORED THEY CAN MAKE US FEEL, NOR ARE WE AWARE THAT STUDIES ALSO SHOW THEY MAKE US HEALTHIER, MORE CREATIVE, MORE EMPATHETIC AND MORE APT TO ENGAGE WITH THE WORLD AND WITH EACH OTHER. NATURE, IT TURNS OUT, IS GOOD FOR CIVILIZATION.'

FLORENCE WILLIAMS,
The Nature Fix: Why Nature Makes Us Happier, Healthier, and More Creative (2017)

THAMES WATER GARDEN, HAMPTON COURT FLOWER SHOW, 1998

Sponsored by Thames Water, this show garden – a collaboration with sculptor Johnny Woodford – had to embrace a 'waterwise' theme: ingenious in the collection of rainwater, frugal with its use and offering drought-tolerant plant associations. I didn't fully appreciate it at the time, but this was a modern-day Paradise garden that proved pivotal to my understanding of why we make gardens and their potency as places of sanctuary. Contained by high walls of reclaimed sleepers carved into waves and a line of burnt-elm spikes punctuating a path around the outside like a tribal redoubt-cum-fortress, the garden was both surreal and challenging to a degree that our Hampton Court Palace backdrop looked a trifle nervous. Aggressive from the outside but peaceful within, the garden retained its function as a place of refuge.

Anyone who has ever built a show garden knows what it's like to live on adrenaline for three weeks during construction and then do your best not to crash while explaining the concept to visitors for six long days. This garden, more than any other, served its purpose to keep us sane with two private spaces tucked just far enough away from the public gaze. We took turns having ten-minute breaks from elbows, criticism and endless questions to be soothed by shade and the sound of babbling water. We also took advantage of a remote-controlled fart machine that was secreted in a rosemary bush at the front of the garden and speculated which of us would be at the other's mercy if we were landed with a royal visit.

As I've already suggested, entering the garden wasn't for the faint-hearted but there was tranquillity within. Hidden butyl liners diverted rainwater to a central pool which was then used to irrigate the garden and feed three phallic water spouts, not unlike instruments of torture you might find in an S & M catalogue (so I've been told). The ribbed phalli (complete with foreskin and razor-sharp teeth) were both beautiful and disturbing, and would have certainly caught the eye of Sir Francis Dashwood (of the Hellfire Club fame), known for the more subtle erotic subtext within the gardens he created in the eighteenth century. I remain to this day disappointed not only that Princess Margaret was completely unfazed by the gruesome gargoyles as Woodford gave her a personal tour of the garden, but also for lacking the courage to press the button on the fart machine when I had the chance.

The Waterwise garden sponsored by Thames Water, Hampton Court Flower Show, 1998, with sculpture by Johnny Woodford

BUPA GARDEN, CHELSEA FLOWER SHOW, 2008

On returning from India in the summer of 2007 with the aforementioned stowaway in my liver, I was asked to design a Chelsea show garden for Bupa that would be relocated to a care home after the show. Health and safety was paramount as the garden would have to be interesting and safe for patients with Alzheimer's and Parkinson's diseases. What with time being short and my ill-health, it was a mighty relief when my initial design was accepted by the sponsor just before I was admitted to hospital. A week or so later, on being discharged and reviewing the design with a clear head, I was horrified to see what I had proposed. A hasty redesign during a half-hour train journey into London (for a meeting with the sponsor before the final submission to the RHS) saved the day. Not my finest moment by a long margin, but thankfully the sponsor understood my reasoning, accepted my honesty and was very happy with the vastly improved and simplified design.

Bupa's recognition and endorsement that gardens are beneficial for mind, body and soul was well documented. It found that gardens not only help stimulate the brain with colour and scent, but also give people a contained world that has a past and present reality, very useful when the rest of a person's life may be becoming confused.

The following benefits were also noted:

- PHYSICAL
  Exercises the eyes through visual scanning, seeing near and far.
  Exercises hands and fingers and upper body.
  Motivates person to walk, stoop, bend, reach, maintain balance.
  Gives pleasure through the senses: seeing, smelling, feeling and hearing (occasional taste too).
  Gives mild to moderate exercise in coordination, strength, stamina and physical activity.
- COGNITIVE
  Enhances orientation.
  Exercises attention span.
  Gives practice in following simple directions.
- SOCIAL
  Promotes interaction by encouraging discussion about common interest.
  Lends itself to many social activities, clubs, garden socials.
- PSYCHOLOGICAL
  Provides a safe, nurturing and familiar environment.
  Provides opportunities to relieve tension, frustration and aggression.

The Bupa garden used a large spherical concrete form (created by artists Serge and Agnès Bottagisio-Decoux) to act as a focal point and destination. Visitors would be drawn into the garden by the 2-metre (7-foot) orb, not unlike the gravitational force of a planet, and given freedom to interact with it while avoiding any sense of a dead end

A garden for Bupa at Chelsea Flower Show, 2008, with sculpture by Agnès and Serge Bottagisio-Decoux

which could potentially confuse anyone with Alzheimer's disease. The circuitous brick path was flat for ease of access, seating was painted red (an aid to recognition for those with Alzheimer's), and a raised bed added a visual level change.

There was, however, a slight disagreement in the choice of plants. I suggested more foliage than flowers to keep the garden calm and allow ease of maintenance. Bupa understandably emphasized the need for colour and scent to stimulate the senses and stir the memory. A compromise was reached, and the front half of the garden was colourful and vibrant while the rear was a verdant oasis of calm.

On reflection, the garden didn't sit well for me. It seemed to lack the overall harmony one always seeks to achieve in a show garden, but thankfully the trees – five multi-stemmed amelanchier – just about held it together. On relocating the garden to its new home in Battersea, the choice of planting turned out to be most appropriate as one half of the courtyard was bathed in sun for much of the day, the other in deep shade. Sadly, a restriction on the number of trees we could use at the care home (a grand total of one) came as a mighty blow. Concerns about how much light the trees would steal from adjacent bedrooms overlooking the courtyard had been expressed, and there was nothing I could say to change their minds. Minutes after a crane lifted the giant boulle into position, along with the solitary tree, a blue tit landed gratefully on a branch of the amelanchier and sang its heart out. As the song reverberated around the courtyard and filtered through open windows wafting joy to a hitherto lifeless space, I could only imagine what five trees would have brought to the new garden.

Raised bed of black basalt for the Bupa garden, Chelsea Flower Show, 2008

TELEGRAPH GARDEN, CHELSEA FLOWER SHOW, 2011

Show gardens are often more difficult to design than real ones. With no existing house and no landmarks or features from which to take reference or inspiration, it's a lonely experience looking at the blank sheet of paper with a deadline looming. Even when you finally arrive at a design you're happy with, the magnitude of the task in hand can seem overwhelming. You have to get everything looking just right on one day in May, and you only have three weeks in which to build it. The fact that there are few restrictions on what can be built sounds liberating, but in reality it's this freedom – this blank page – that often proves to be the biggest hurdle. I'm not always clear where ideas for show gardens come from, but in this case the mysterious workings of the subconscious had much to do with the final outcome.

From the outset I was determined to use the sculpted concrete columns ('Les Colonnes') made by Serge and Agnès Bottagisio-Decoux. I had seen and fallen in love with them when I first visited the artists' residence and workshop in 2005. I'd been reluctant to use the forms in earlier show gardens for fear of over-egging the omelet,[4] but this time I would design the garden with these beautiful forms in mind.

A month or so after submitting my design – a contemporary sunken gravel garden loosely inspired by the Roman ruins of Ptolemais in Libya, which I'd experienced a year or two before the fall of Colonel Gaddafi – I chanced on a photograph of my mother seated on a block of stone at the ruins. It hadn't crossed my mind at all during the design process, but this scene clearly played a significant part in it. The fact that my mother had only recently died added fuel to a slightly crazy

Mixed perennials for the *Telegraph* garden, Chelsea Flower Show, 2011

notion that I was subconsciously building a private memorial to her at the Chelsea Flower Show. I've no doubt she would have loved the idea, but still raw from the sudden and untimely nature of her departure, I kept this part of the story to myself, largely for self-preservation and also to avoid anyone thinking I was looking for a sympathy vote when it came to judging. Care was taken to find a piece of stone that was almost identical to the one on which she sat among the ruins. Naturally, I was sad she wasn't around to share the garden's success, but as a cathartic exercise in helping me come to terms with her passing, that in itself was worth more than all the accolades the garden received.

Sunken gravel garden for the *Telegraph*, Chelsea Flower Show, 2011, with sculpture by Agnès and Serge Bottagisio-Decoux

HORATIO'S GARDEN, SALISBURY HOSPITAL, 2012

In 2011 Olivia Chappell asked me to design a garden to celebrate the life of her son Horatio. He was involved in a fatal accident during a school trip when their camp in Svalbard was attacked by a polar bear. Horatio had hoped to follow his father's footsteps in the medical profession and had canvassed patients at the spinal unit where his father worked as a surgeon to find out how their stay at the hospital could be improved. Top of the list was safe access to a garden or outdoor space.

What Olivia didn't know was that I was already familiar with the spinal unit. A family member had spent a year there being rehabilitated after a serious diving accident that left him paralysed from the waist down. I remembered a grubby smokers' corner and a gloomy playground virtually inaccessible for any but the most confident and adventurous wheelchair user. Naturally, I couldn't refuse.

Questionnaires completed by patients about what they wanted from the garden were typical in terms of colour, scent, texture and shade, so this was fairly straightforward. However, an early idea of sloping the paths down to a central courtyard to accentuate the height of plants without having to resort to raised beds was met with much resistance. Being wheeled around in a bed and a chair made me appreciate how vulnerable one feels with someone else steering (you can feel every hump), but it didn't make me appreciate just how challenging it is to negotiate slopes. The idea was quickly binned.

As a substitute for my personal dislike of raised beds, an idea for spine shaped drystone walls was tabled as a structural feature. The walls, dissected by a path leading to a complete spine, drew heavily on the experience of patients being rehabilitated, the metaphor being that patients arrive at the unit broken but leave mended spiritually if not physically. While this may not always be the case, and some people will have differing opinions on how their path to recovery has been handled, the feedback we get from patients using the garden is clear: time spent recovering from their traumatic injuries has been made more bearable.

Having to compete with the overbearing architecture of the hospital and adjacent car park meant aligning the axis of the garden to make the most of a view to the Clarendon Way. We were clear from the outset that this shouldn't be a low maintenance garden. A perennial-heavy design would mean more maintenance, but the benefits of a more dynamic landscape in terms of wildlife, colour, scent and sound would heighten the overall impact of the garden and provide a visual feast for creative exercises such as art therapy, poetry and wildlife studies. It would also accentuate the seasons, giving patients something to look forward to and encouraging visitors back at different times of the year for fund-raising events like food fairs, plant sales, National Gardens Scheme (NGS) open days and musical events. These not only help future-proof the cost of ongoing maintenance of the garden, but also provide both seasonal

LEFT Horatio's Garden with sculpture by Johnny Woodford BELOW Mixed perennials celebrate the seasons and attract a range of insects which, in turn, attract birds into the garden.

and cultural interest for patients, some of whom have to endure long stays (one year or more) before being rehabilitated.

Horatio's Garden is large enough to accommodate a number of spaces (all accessible via generous resin-bonded paths) where patients and family can find important privacy. A large open area has proved useful for events enjoyed by a hundred or more people, and a conventional space near a gymnasium is more or less dedicated to occupational therapy where patients can improve their motor skills in sowing seeds and potting up plants. As evidence continues to be updated and processed, designers can now work with clinicians to tailor their designs to address the specific needs of patients and medical teams to get the very best from a given site.

Since the garden was completed in 2012, we've seen at first hand how access to a garden improves the well-being of patients with life-changing injuries. We've seen how it affects other family members and also the dedicated staff who work hard to heal patients mentally and physically and, at the very least, come to terms with their situation. We've seen how plants encourage insects and birds, accentuating the overall experience of providing distraction, privacy, entertainment and knowledge about the natural world, for not everyone coming to the hospital is a gardener. We've also seen how communication between doctors and patients has been improved through the vital work of volunteers looking after the garden. Once allowed outside, patients

using the garden are generally more relaxed than they are indoors, often engaging with volunteers and occasionally imparting information that they may not have been able to express in the sterility of the hospital ward. Unforeseen, this dynamic has been welcomed by care teams and patients alike.

The importance of maintaining a garden can't be overemphasized. It's often the first conversation I have with clients these days, especially if it's for a school or hospital. A series of events overseen by a creative team of organizers and administrators means that the future of Horatio's Garden is assured.

At the time of writing, four out of the eleven spinal injury centres in the UK have a Horatio's Garden, with two in the pipeline. This is down to the extraordinary vision and commitment of Horatio's mother, Olivia. She and her family have turned tragedy into something positive and beneficial to so many people whose lives have seen dramatic upheaval.

WE´VE SEEN HOW PLANTS ENCOURAGE INSECTS AND BIRDS, ACCENTUATING THE OVERALL EXPERIENCE FOR PATIENTS BY PROVIDING DISTRACTION, PRIVACY, ENTERTAINMENT AND KNOWLEDGE ABOUT THE NATURAL WORLD.

Paths are wide enough to allow patients to enjoy and interact with the garden even from their beds.

M&G GARDEN, CHELSEA FLOWER SHOW, 2016

An introduction to a film about forests said, 'Carry the forest within you and you'll always be home.' This is precisely what I was trying to express with the M&G garden, albeit within a very contrived and contemporary framework: a conceptual representation of 'home is where the heart is.' Perhaps even more self-indulgent than the *Telegraph* garden, this was an opportunity to reflect on part of my youth that played an important part in my eventual decision to become a landscape designer. In a way, this proved something I've always suspected: that gardens tend to be more powerful when created by the owner. I knew this garden, or at least its intended atmosphere, before I'd even created it. It belonged to me.

Of all natural landscapes, woodland is perhaps the most widely appreciated. There's a primeval connection with natural forests that's hard to ignore, and they are wonderful antidotes to the stresses of our technological world. A walk in a forest helps us reconnect with nature. It calms and soothes. Studies have shown they can boost the immune system and accelerate recovery by increasing red blood cells, improving sleep and reducing blood pressure. This can mean a reduction in the need for medication and can help centre oneself amidst the distraction of decision making and information overload.

As the garden was based on childhood memories and feelings that had to be dredged by a net forty years long, there was always a risk that these personal associations might either alienate the visiting

The M&G garden at the Chelsea Flower Show, 2016, inspired by early memories of Exmoor National Park

public or complicate the brief and the overall story surrounding the garden. One such element was a private homage to David Bowie, who had been a huge influence during my teenage years. The initial idea was to make a black star in the base of a pool at the centre of the garden as a reference to Bowie's last album. Worried about it being too literal, stealing the limelight from the pool itself and bordering on the sort of cheesiness I moan about in other gardens, I googled 'Black, Starfish, Porlock' in case there was a more subtle avenue to explore. To my astonishment and delight, there is a rare brittle starfish that has occasionally been found in Porlock Bay and, yes, you've guessed it, it's black. Some things are meant to be.

The interesting thing is that it was the least stressful of all the gardens I've built because, in an odd sense, it felt like home. All I can say is that I was extremely fortunate and grateful to have had the opportunity to explore that notion. I never really like saying 'never again', but it did feel like the end of a journey and that there was little else to say in the world of show gardens. Only a vegan-inspired garden will tempt me back.

CHRIST CHURCH SCHOOL, LONDON, 2019

In 2016 I had the good fortune to be introduced to pupils at Christ Church CE Primary School in Battersea, south London. The children start gardening from the minute they arrive and have the use of a private walled garden nearby to grow vegetables. Teachers noticed that children respond more positively outdoors and use the garden as a valuable resource for learning in all subjects, and studies have shown that outdoor learning concentrates the mind and can act as a sort of decompression chamber or safety valve in which to restore calm if things get out of hand. A garden of their own within the school grounds has always been their objective, so I was delighted to be in a position to help them transform a small patch of land adjacent to their playground. What comes across very clearly is the respect children have for nature and the creatures that share the garden with them. Biodiversity is encouraged and the children are taught that all living beings in the garden should be respected. Nothing is killed deliberately. This refreshing approach engenders a positive relationship with the natural world and may help them develop a more compassionate approach to life at the school and beyond.

The school isn't completely risk averse, believing that careful planning and coordination can help children develop an understanding of the hazards associated with the natural environment. This imparts a sense of taking responsibility while enjoying green spaces where they can learn about the environmental, social, economic and health benefits that these spaces offer. Activities in the garden have a positive effect on well-being, physical strength and cardiovascular fitness, which in turn can improve mood, social interaction, concentration and emotional stability. They also offer unique opportunities to improve confidence and self-esteem. The simple but magical act of growing something from seed, for example,

Pupils from Christ Church CE Primary School, Battersea, London, visiting our allotment

can be enormously satisfying and life-affirming, which can boost confidence and create a powerful connection with and understanding of the natural world. Such a grounding in respect for all life from such an early age won't lead to world peace overnight, but reinforcing the notion of compassion must surely be one the more important traits to impart on the young if we want a healthy, functional society.

Whether the children maintain an interest in gardening during their teenage years is another matter, but the fact that they will leave primary school knowing the importance of maintaining and improving biodiversity, respect for all life forms and how to grow their own food will give them a good platform to come back to. With the prospect of peak oil (the point at which the maximum rate of crude oil extraction is reached) and the knock-on effects of fuel shortages reinforcing the need for locally grown food, such skills shouldn't be underestimated.

MAGGIE'S CENTRE, CARDIFF, 2019

Maggie's Centres create visionary and secure environments for people affected by cancer. They were the brainchild of the late Maggie Keswick, who, in 1993, learned that her breast cancer had returned and that she had three months to live. Through an advanced chemotherapy trial she lived for another eighteenth months, and in that time worked with her husband, Charles Jencks, and oncology nurse Laura Lee to develop a new approach to cancer care. Her hope was that people should be more informed about their treatment and given free resources in a calm, friendly environment with access to support in terms of stress management and psychological advice. Maggie's Centres enable patients to make educated choices about their options during a vulnerable time of their life and above all, in worst case scenarios, 'not to lose the joy of living in the fear of dying'. Key to this was a place where all this could function with the minimum of fuss and where you could talk to others in similar situations.

Maggie's Centres are much celebrated for their visionary architecture. While imaginative and liberating, they are also homely and welcoming. In most cases, the first thing a visitor will encounter is a kitchen where someone will offer you a cup of tea. This initial experience of walking into a building and feeling 'at home' is important. We all know the less than friendly and often disorientating corridors and waiting rooms of hospitals where privacy, information and even a view are hard to find. Being able to see trees, plants, wildlife and sky serve both as a distraction and a means of maintaining a connection with the real world at a time of confusion and stress. Accessing the view is a real bonus.

In 2012 a large site at the Velindre Hospital in Cardiff was identified within an abandoned Victorian garden adjacent to a wooded nature reserve. This proved too ambitious for the available purse strings so, at the time of writing, work has started on an interim site with a building designed by Dow Jones Architects. Maggie's Interim is a modest facility with a small courtyard at the entrance and a thin strip of land to the rear bordering a rather uninspiring screen of Leyland cypress trees, a cherry and some dead English elm trees.

To be honest my heart sank when I saw the cleared site and the limited scope, but the architects saved the day by obtaining permission to thin some of the lower branches of the leylandii in order to allow vignettes and provide a little light for some woodland-edge planting along the boundary. It's not a space anyone can access, but some of the most alluring spaces we experience are the ones which we can't physically get to. However, not being able to immerse oneself among plants in the rear garden might be a source of frustration to anyone with a conventional idea of how a garden should work. The intention is that the borrowed space beyond the ha-ha serves as a backdrop to the building that visitors to the centre can enjoy both from inside and out along the adjacent path. Skeletal stems of *Aralia elata* planted on a bank of ivy will tease views to twisted, hulking forms of conifer trunks that animate the space, not unlike a Peter Doig painting or some David

Nash tree art. 'Nests' made from dead twigs and random logs that punctuate the space will in time be consumed by ivy to create habitat mounds. Random elm and cherry will come and go, some left to decay with dignity. Brambles will fight dog rose, honeysuckle and clematis for the upper hand, while foxgloves, ferns and tufted hair grass will accentuate the notion of woodland edge. The hope is that visitors will find beauty not just in the seasonal notes that manage to get a foothold in this challenging patch of ground, but also in the less familiar tangle of nature and the often invisible life it sustains.

Amelanchier lamarckii in the courtyard at the entrance to Maggie's Cantre, Cardiff

4 THE ENVIRONMENT

‘1 KG OF INTENSIVELY REARED BEEF REQUIRES UP TO 10 KG OF ANIMAL FEED AND 15,500 LITRES OF WATER. IT PRODUCES AS MUCH POLLUTION AS DRIVING FOR THREE HOURS WHILE LEAVING THE LIGHTS SWITCHED ON AT HOME.’

FRIENDS OF THE EARTH, ‘WHAT’S FEEDING OUR FOOD? THE ENVIRONMENTAL AND SOCIAL IMPACTS OF THE LIVESTOCK SECTOR’ (DECEMBER 2008)

Killed at Newman's Abattoir on 17 April 2019

The rise to fame of Greta Thunberg, a sixteen-year-old schoolgirl from Sweden, from her 2018 'school strike for climate' outside the Swedish parliament to her address to the United Nations in New York in September 2019, has been instrumental in moving climate change up the political agenda around the world. She has spoken to the public and politicians throughout Europe and in the US and in May 2019 was featured on the cover of *Time* magazine. Shortly before, during the Easter break, she met with political party leaders in London to discuss her concerns about our addiction to fossil fuels and the future well-being of our planet. Her visit to the UK coincided with two weeks of protests organized by climate change activists Extinction Rebellion (ER) which caused widespread disruption in London and other cities around the world. Bridges, roads and financial institutions were blocked by thousands of protestors raising awareness about climate breakdown and the government's complete failure to take the issue seriously. While the British press focused on the inconvenience of travelling in central London and the amount of money that businesses were losing, the public took time to listen and understood that these minor inconveniences were nothing compared to the major changes coming to our lives if no concerted effort on a global scale was made to reduce carbon emissions. With 1,000 ER protestors arrested and David Attenborough pulling no punches on the BBC's screening of *Climate Change: The Facts* it served as a pivotal fortnight of awakening.

Exactly why it's taken so long is anyone's guess, but a mixture of politics, denial and the public's overriding resistance to change has much to do with it. The things we have been taught since childhood and cultural conditioning mean that questioning what we've always thought of as 'normal' is challenging, even to those who consider themselves open-minded, liberal or progressive.

A few weeks later a report by the Intergovernmental Science-Policy Platform on Biodiversity and Ecosystem Services (IPBES) confirmed: 'Biodiversity – the diversity within species, between species and of eco-systems – is declining faster than at any time in human history.'[2]

ANTHROPOCENE: THE CURRENT GEOLOGICAL AGE WHERE HUMAN ACTIVITY IS THE DOMINANT INFLUENCE ON CLIMATE AND THE ENVIRONMENT.[1]

It's hard to imagine a more stark warning. It's also unfortunate but hardly surprising that the predicted collapse of life on earth as we know it, the biggest and most disturbing news that humanity has ever known, was upstaged by a royal birth.[3] Even more disturbing was the lack of engagement by the gardening press apart from token tips on Twitter on how to attract wildlife to your garden.

The world of horticulture has acknowledged climate change for many years, albeit with an air of inevitability rather than any sense of alarm. Magazines and books look at how we can adapt to the predicted changes and even profit from them. Few, if any, want to examine why it's happening and what we can do to avert it. My letters to both the RHS and Royal Botanic Gardens Kew questioning their position on climate change were met with a telling silence; conversations I had with journalists in the garden media were awkward and frustrating. Just what does it take to acknowledge that for all the good we derive from gardens and gardening, the largest and most important garden of all – our Garden of Eden, planet Earth – is threatened by our very existence, a species that accounts for just 0.01 per cent of all life?[4]

'MOST PEOPLE CLAIM THEY WILL DO "ANYTHING" FOR THEIR CHILDREN, UNTIL YOU INFORM THEM THAT EATING ANIMAL PRODUCTS IS DESTROYING THE PLANET THAT THEY WILL GROW UP ON.'

CONNOR ANDERSON,
@C_Anderson1998, Twitter, 28 January 2019

Our demand for resources is unreasonable in the extreme and the notion of perpetual growth unsustainable.[5] We understand more clearly than ever the impact of human intervention and the destabilizing effect it's having on the earth's natural cycles. Rainforest destruction, species extinction, ocean depletion, ocean dead zones, loss of biodiversity, global warming, pollution of air, land and waterways, and water shortages are just some of the more pressing problems we face, but there are two reasons why they are almost instantly forgettable. First, it's a case of 'out of sight, out of mind' as many of these issues aren't (yet) happening on our doorsteps. Second, one of the leading causes of all these problems is, of course, animal agriculture. What's frustrating is that it doesn't have to be that way, as many of these problems can be mitigated by humans adopting a plant-based diet. It may seem challenging, daunting even, but it is in fact the easiest way to reduce our carbon footprint and the negative impact we have on a whole range of environmental problems.[6] As an intelligent species with a sense of moral agency, we could argue that collectively we have an ethical duty to preserve what the planet provides for future generations of all life forms, especially those that we depend upon for our very survival. Gardeners, therefore, with one finger on the pulse of mother nature, should be at the very helm of the vegan movement with a genuine concern for a more sustainable future.

But while such a concern for the environment seems to be inherent in most gardeners, going the extra mile and adopting a plant-based diet can seem like a step too far. Even in May 2019, when the UK Parliament finally accepted that there is indeed a climate emergency, suggestions as to what the public can do about it fell short of acknowledging advice from UN scientists and researchers at the University of Oxford that cutting meat and dairy from our diets can massively reduce each individual's carbon footprint and help mitigate many other impacts on the environment. Animal agriculture is perhaps the biggest elephant in the room that you will ever (not) see.

We've become so conditioned when it comes to exploiting animals that even some of the most open-minded and compassionate people will not entertain the notion of giving up eating them to preserve the

Belching livestock account for 18 per cent of greenhouse gas emissions.

future of life on earth as we know it. The love of eating meat helps fuel a delusion that we are the most important species on the planet and that climate change will always be someone else's problem. Our only hope is for a paradigm shift among those with a genuine concern for the world we enjoy, not to mention the world we want to leave for our grandchildren, and that they will be selfless enough to see beyond what we perceive as 'needs' and take action. We need to understand that nature is essential to our survival. For example, 75 per cent of global food crop types, including fruits and vegetables, and some of the most important cash crops, such as coffee, cocoa and almonds, rely on animal pollination; marine and terrestrial ecosystems are the sole sinks for anthropogenic carbon emissions; loss of coastal habitats and coral reefs reduces coastal protection, increasing risk from floods and hurricanes to life and property for 100 million to 300 million people living within coastal flood zones. Clearly, a more compassionate and altruistic outlook will help us sustain not only ourselves as a species but, more importantly, all the other life forms that make our Garden of Eden habitable.

CLIMATE CHANGE, LAND AND GRASS-FED CATTLE

It was a revelation to me that belching livestock account for 18 per cent of greenhouse gas emissions (GGE) (including methane, which has twenty-three times the global warming potency of carbon dioxide), more than the world's entire transport sector,[7] and that embracing veganism is the single most effective way of reducing our environmental impact. The aforementioned five-year study by researchers at Oxford University concluded that removing animal products (meat, fish, dairy and eggs) from your diet could reduce your greenhouse gas footprint from food by up to 73 per cent and would reduce the amount of farmland needed by up to 75 per cent, an area of land equivalent to the USA, China, Australia and the European Union combined.[8] This would have a significant effect in reducing GGE and would free up land lost to agriculture, one of the main causes of species extinction.

The study, one of the most comprehensive to date, collected data from 38,000 farms in 119 countries, and found that the meat and dairy industries are responsible for 60 per cent of agriculture's GGE while providing only 18 per cent of calories and 37 per cent of protein consumed around the world. That, together with other negative impacts such as excessive land and water use, eutrophication[9] and global acidification, means that a vegan diet has a much more positive effect across a wider range of environmental impacts than simply cutting down on flights or buying an electric car.

THE MEAT AND DAIRY INDUSTRIES ARE A LEADING CAUSE OF GLOBAL WARMING

- While transport and energy sectors are responsible for much of the carbon dioxide emissions, animal agriculture is responsible for 53 per cent of nitrous oxide, which has 296 times the potency of carbon dioxide, and 44 per cent of methane, which has thirty-four times the warming potential of carbon dioxide.[10] Worryingly, methane emissions from livestock have been estimated to increase by 60 per cent by 2030: the same time period we need to see strong and rapid reductions of all greenhouse gases.[11]
- Methane is responsible for 24 per cent of today's warming, further demonstrating the potency of this greenhouse gas, despite being released in much smaller quantities compared to carbon dioxide.[12]
- Energy-related carbon dioxide emissions are set to rise by 20 per cent by 2040. Emissions for agriculture are set to rise by 80 per cent by 2050.[13]

Naturally there has been pushback from the meat and dairy industries suggesting that grazing livestock, in particular grass-fed cows, can remove carbon dioxide from the atmosphere,[14] but the results of a comprehensive study by the Food Climate Research Network (2017) suggest otherwise. The report, entitled 'Grazed and Confused', explains that 80 per cent of livestock emissions comes from cows and sheep and, of this, only 20–50 per cent of their emissions can be offset: 'In some circumstances, you can get carbon capture, but not always and the effect is small. You cannot extrapolate from a nicely run Dorset farm to a global food strategy . . . Grazing systems and grass-fed beef may offer benefits in these respects, benefits that will vary by context. But when it comes to climate change, people shouldn't assume that their grass-fed steak is a climate change-free lunch. It isn't.'

Trying to navigate a way through claims and counter-claims is enough to drive a teetotaller to drink, but to summarize:

'LIVESTOCK IMPACTS ON ECOSYSTEM GOODS AND SERVICES ARE LARGELY NEGATIVE, THROUGH IMPACTS SUCH AS DEFORESTATION, NUTRIENT OVERLOADING, GREENHOUSE GAS EMISSIONS, NUTRIENT DEPLETION OF GRAZING AREAS, DRYLAND DEGRADATION FROM OVERGRAZING, DUST FORMATION, AND BUSH ENCROACHMENT.'

United Nations Millennium Ecosystem Assessment 2005, taken from Friends of the Earth, 'What's feeding our food? The environmental and social impacts of the livestock sector' (December 2008)

- Industrialized farming contributes massively to GGE and the food grown to feed these animals causes deforestation, species extinction and pollution of waterways on a massive scale. Monocultural arable crops like soya and corn or maize (the meal of which is fed to animals after oil extraction) are highly damaging and unsustainable in the long run.
- Grass-fed livestock are better but not ideal. People should be seriously cutting down on their meat consumption for similar reasons, as deforestation to make way for more pasture has the same negative consequences.[15] A shift from industrial farming to grass-fed beef would be better than nothing, but such meat would almost certainly be unaffordable to the majority of people, who consume heavily subsidized animal products from intensive farms.
- Producing grass-fed meat is an extremely inefficient use of land, using twice as much land globally as for crop production[16] and only producing 1.2 per cent of our protein.[17] On a global scale 33 per cent (471 million hectares) of arable land is used for feed crops.[18]
- The UK has 8.4 million hectares of permanent pasture and 5.8 million hectares of cropland, of which 55 per cent is currently used to produce feed for farm animals. Overall, 48 per cent of land in the UK is used for animal agriculture – either for pasture or feed crop production.[19]
- Land dedicated to producing cereals for human consumption delivers about ten times more protein than it would if it were used for rearing animals.[20] Even if grass-fed meat was the silver bullet for non-vegans, there just isn't enough land in the UK or the rest of the world[21] to produce enough of it and certainly not at a price that most people could afford.[22]
- With almost half of all land in the UK currently being used for farming animals, repurposing it represents a good opportunity to meet climate goals, as it provides very little nutrition compared to the resource inputs involved. It will therefore be necessary to reforest some land in the UK.[23] Research from Harvard University (2019) shows that the UK would be able to sustain itself and meet its Paris Climate Agreement obligations (to reduce GGE by 80 per cent by 2050) by returning land used for animal agriculture back to forest.[24] Reforesting land used for grazing and feed crops could soak up twelve years of carbon dioxide emissions and bring

a number of benefits such as water filtration, flood control, soil health, more habitat for wildlife and more forests for recreation. In an alternative scenario, the repurposing of just over half of cropland used for animal feed could maximize food self-sufficiency in the UK with increased and more diverse fruit, pulse and vegetable production for human consumption.

- A review in the journal *Agricultural Systems* says that there is no scientific evidence to support Intensive Rotational Grazing (IRG), a technique which has garnered support from the grass-fed meat community with claims that it can reverse desertification, store carbon and, by inference, offset climate change. 'The vast majority of experimental evidence does not support claims of enhanced ecological benefits in IRG compared to other grazing strategies, including the capacity to increase storage of soil organic carbon . . . IRG has been rigorously evaluated, primarily in the US, by numerous investigators at multiple locations and in a wide range of precipitation zones over a period of several decades. Collectively, these experimental results clearly indicate that IRG does not increase plant or animal production, or improve plant community composition, or benefit soil surface hydrology compared to other grazing strategies.'[25]
- Government funded research and experimentation are needed to find better ways of producing food with an emphasis on carbon-building crop rotations, perennial crops, agroforestry, cover crops and intercropping to protect and improve microbial health of the soil. Not even organic gardening has had the full attention it deserves, so it will take time for new forms of agriculture (veganic gardening) to emerge and for farmers to transition to arable.
- Claims that grass-fed systems increase biodiversity aren't strictly true. Jenny Hall, co-author of *Growing Green* (2006), explains: 'Most commercial grasses, that have nitrogen and hormonal/chemical herbicides added, have no value for biodiversity. Adding nitrogen wrecks that. Wildflowers require the running down of nitrogen and most hay cuts for industry do not allow wildflowers to set seed.'[26] Anyone who has tried to grow a wildflower meadow will know just how difficult it is, the starting point being to reduce fertility by removing clippings and even stripping off topsoil. Pasture fit for beef and dairy relies on nitrogen-rich soil, and this, together with phosphorus and potassium, is added to boost fertility for a hungry herd.[27] The idea that cows roam around in fields full of wildflowers is a romantic notion, but successful management of species-rich meadow pasture is quite an art and timing is everything. Grazing has to be carefully controlled and land has to be rested to recover, so it's a costly process.
- Arguments about wildlife displacement and destruction of habitat in arable farming don't usually take into account feed conversion ratios in animal agriculture. Livestock have to be fed for months before being slaughtered; therefore the feed conversion ratio (a comparison between the amount of food necessary during their life and the amount of food you get from their carcass) is always negative. The simple fact is that more crops and land are required to keep animals, and this means more wildlife displacement than arable farming.[28]

Upland sheep farming, in particular, causes erosion and compaction that accelerate water run-off which, in turn, causes landslips and flash floods, thus reducing fertility and exacerbating drought in more productive lowlands.

- Soil conservation measures need to be improved on arable land with more hedgerows, trees and green manure that's allowed to flower.
- There needs to be a shift towards more soft fruit, orchards, nut groves, agroforestry and forest gardening to make more efficient use of arable land: more people nourished per hectare.
- Freed up land unsuitable for arable crops could be used for rewilding and the restoration of forests and degraded wetlands.
- Results from research sponsored by the meat and dairy industries tend to be heavily biased towards the products they are making money from. Independent studies by scientists with no financial interest tend to lean towards the fact that plant-based diets are better for the environment and personal health.
- 156 billion land animals are killed each year globally for food.[29] Factory farming takes up more space when you take into account the land it needs to grow the grains for feed. If we fed these grains to humans instead of animals, we could feed an extra four billion people.[30]
- A study in the US shows that changing from feed crops for animals to health-promoting crops for humans could feed twice as many people from the same area of land.[31]
- Studies show that ploughing releases carbon into the atmosphere and that soil microbes play a crucial role in the sequestration of carbon, so trials in no-till and other veganic methods are crucial if we are to find a reliable and sustainable way of using land for arable crops.
- Even if a well-managed arable system becomes the go-to method for future food security, it will take some time (with incentives, skills and training) to transition to plant-based agriculture.

SOYA VERSUS MEAT

The benefits of soya as a useful source of protein have attracted some unfair criticism from non-vegans looking for loopholes and arguing the case for eating meat. They will tell you that the vegan lust for soya is causing deforestation and that a shift to a plant-based diet will only serve to increase the demand for soya worldwide. That's not strictly true. Soya bean meal is a byproduct of the seed oil industry and most of this (80 per cent of the world's soya production) is used to make animal feed, so taking this into account, not to mention the fact that soya is added to many non-vegan processed foods and ready meals, the demand for soya in a vegan world would actually go down. Furthermore, animals are extremely inefficient in turning that food into meat: on average, they need around 8 g of plant protein to create 1 g of meat protein. It's far more efficient for us to eat the protein straight from the original plant source, so by eating soya instead of meat the worldwide demand for soya in a vegan world would go down.[32] But we can do even better than that if we all (vegans included) cut down on the use of margarine spreads and other processed food where soya oil is a key ingredient.

AMOUNT OF EMBEDDED SOYA PER 100 G OF ANIMAL PROTEIN

- Chicken – 109 g (more than the weight of the actual meat)
- Farmed salmon – 59 g
- Eggs – 64 g
- Pork chops – 51 g
- Hamburgers – 46 g
- Cheese – 25 g

Source: WWF, 'Average EU citizen consumes 61 kg of soy per year, most from soy embedded in meat, dairy, eggs and fish' (18 May 2015), www.panda.org.

RAINFOREST DESTRUCTION AND SPECIES EXTINCTION

While animal agriculture is one of the biggest drivers of climate change, its effect on other environmental issues is no less disturbing. Until I became vegan my understanding was that the rainforest was being cut down mainly for timber. It turns out that up to 80 per cent of the Amazon rainforest destruction these days is attributed to animal agriculture through land cleared for grazing or growing grain to produce vegetable oils and to feed livestock.[33] Greenhouse gas emissions from the burning of cleared forests further accentuate the negative effects on climate change, not to mention habitat destruction, loss of biodiversity and species extinction.

Since 2008, in an attempt to limit the damage to Amazon rainforests, Brazil has instead exploited the tropical savannah region of Cerrado, an important ecosystem that supports 5 per cent of all species on the planet. An area larger than South Korea has already been lost to animal agriculture, and this has contributed significantly to GGE and climate change. It's also put an enormous strain on water resources; the vanishing streams and erosion are having a detrimental effect on wildlife and plants are disappearing rapidly. From the farmers' point of view, it's a sacrifice worth making as Brazil sees itself as a vital contributor to food security. Even more worrying is the fact that Brazil's president, Jair Balsonaro, has transferred the regulation and creation of new indigenous reserves in the Amazon to the agricultural ministry, paving the way for an increase in deforestation and violence towards indigenous people. Balsonaro argues that developed counties, who cut their forests down years ago,

ACCORDING TO THE WORLD HEALTH ORGANIZATION (WHO) AND THE FOOD AND AGRICULTURE ORGANIZATION OF THE UNITED NATIONS (FAO), 'THE NUMBER OF PEOPLE FED IN A YEAR PER HECTARE RANGES FROM 22 FOR POTATOES AND 19 FOR RICE DOWN TO 1 AND 2 PEOPLE RESPECTIVELY FOR BEEF AND LAMB.'

WHO, 'Nutrition – Global and regional food consumption patterns and trends: 3.4, Availability and changes in consumption of animal products', www.who.int/nutrition

should pay if they want Brazil to save the Amazon. This doesn't strike me as unreasonable, but with the USA and Brazil drawing up agreements to promote private-sector development in the Amazon, the world's nations need to act fast to avert such deals and work with Brazil to let the indigenous tribes steward the forests as they have for millennia.[34]

And this is not just a problem in the developing world. Queensland in northern Australia has been a deforestation hotspot for the past fifty years due to clearance of native forest for pasture and feed production. The toll on wildlife might not be apparent for decades, but in just over two years the WWF estimated 68 million animals were killed as a direct result of tree cutting. It's also had devastating effects on the Great Barrier Reef, which needs clean water to recover from bleaching events (due to water temperature rise), and this has been greatly impeded by sediment washing in from deforested land and chemicals washed in from feed crop land.[35]

Just before I completed the final draft of this chapter, a study in the journal *Science Advances* revealed that GGE caused by the destruction of rainforests around the world have been underestimated by a factor of six.[36] It turns out that when GGE are declared they don't account for the carbon dioxide that would have been soaked up by the forests (known as 'forgone removal') if they'd been left alone. The figures are said to be conservative as the study didn't include emissions from other woodlands and the great boreal forests in high latitude environments. It stands to reason that if we are losing forests we are losing habitats, and if we're losing habitats we're losing biodiversity.[37] It's a domino effect with dire consequences. The fact that we're currently in the throes of a sixth mass extinction event[38] should ring as many alarm bells (if not more) than climate change.

SOIL AND BIODIVERSITY

Whether the future of food security is vegan or not, one of the most important and often overlooked issues lies under our very feet: the soil. Crops can't be harvested continuously without replacing the nutrients and organic matter that are lost in the process.

Bacteria, fungi and other microbes play a vital part in breaking down organic matter, fixing nitrogen and making nutrients available to plants.[39] An acre of fertile land contains around 4 tonnes of micro-organisms which keep soil and plants healthy and disease-resistant. Despite the pioneering work of Lady Eve Balfour (founder of the Soil Association and author of *The Living Soil*, 1943), microbial ecology and its effect on yields and sustainability, not to mention how ecological and biochemical cycles might affect climate change, are only just getting the attention they deserve.

Iain Tolhurst's veganic techniques as a vegetable grower, a closed system where no animal inputs are used, are an important template for farmers looking to transition from livestock to arable (see Chapter 2). His system uses plant diversity to boost biological activity in the soil, which in turn improves the availability of nutrients to plants while reducing dependency on fossil fuels. Green manures (cover crops) and the recycling of organic waste build soil structure, improve drainage and water holding capacity, and keep sustainable levels of nitrogen in the soil. The addition of flowering annuals also enriches biodiversity, something that is seriously lacking on intensive arable farms where pesticides and herbicides are still used and where 25–40 billion tonnes of topsoil are lost each year to erosion caused by ploughing and intensive cropping.[40]

Carbon is also lost to the atmosphere through ploughing, and this is a significant contributing factor to climate change.[41] Curiously, this is occasionally used as a loophole by non-vegans advocating grass-fed meat to debunk the idea of a vegan world. Aside from being dangerously elitist, the idea that people should reduce the amount of meat they eat and consume only organic, grass-fed steak completely ignores everything else they should be eating for a balanced diet. If you are advocating a diet that doesn't involve industrial farming (and let's not forget that even in high-welfare systems, non-ruminants such as pigs, chickens and farmed fish rely heavily on feed), the vast majority of this 'balanced' diet is vegan. Blaming vegans, therefore, for the amount of carbon lost to the atmosphere through ploughing is disingenuous. Are we to assume that no one, aside from the small number of vegans in the world, eats plants? Are we to assume that the carbon lost to the atmosphere from fields that are ploughed to grow food for animals (that could feed an extra four billion people) eaten by non-vegans should be blamed on vegans? Even without no-till, organic systems, if the world went vegan overnight the amount of ploughing wouldn't need to increase.

A transition to a veganic system, therefore, would reduce both the amount of arable crops grown and the loss of carbon. Mycorrhizal fungi, which improve the nutrient uptake of plants they feed from, also help to store carbon in the soil, so the adoption of a veganic system with no-till and a diverse range of plants can only improve both soil fertility and carbon storage.

Land considered unsuitable for arable farming has traditionally been used for livestock, but this is to the detriment of native systems (including forests) and the wild animals that might inhabit them. According to

George Monbiot addressing journalists at the Animal Rebellion occupation of Smithfield Market, London, on 7 October 2019

the journal *Science Direct*, upland sheep farming, in particular, causes erosion and compaction that accelerate water run-off which, in turn, causes landslips and flash floods, thus reducing fertility and exacerbating drought in more productive lowlands.[42] A memorandum from DEFRA states: 'There are a range of environmental effects from overgrazing that impact upon wildlife, soil structure, water quality and carbon emissions. Birds can suffer from reductions in availability of nesting sites for those species preferring higher sward levels for nesting. Ground nests are more vulnerable to trampling when high stock densities occur. Reductions in food sources can also become a problem in areas of high grazing. Small mammals may lose ground cover increasing their risk of predation.'[43] Rewilding such land and the natural water catchment would allow trees and shrubs to stabilize the soil and hold water like a sponge, releasing it more gradually downstream.

Knepp Farm, a 1,416-hectare (3,500-acre) estate in Sussex, is a stunning example of how rewilding can breathe life back into land considered unproductive. It once languished as an unprofitable wheat and dairy farm, but then the owners released cattle, pigs, deer and Exmoor ponies and left them to their own devices. The result was a resounding success with restored ecologies boasting marshes, new waterways and some of the highest numbers of nightingales, turtledoves and purple emperor butterflies in the UK. With the combination of eco-tourism, glamping and 'high-quality' meat, the farm is now profitable, and the success of this rewilding project gives us reason to be optimistic about how the landscape might change in a vegan world with improved biodiversity, cleaner air and waterways.[44]

George Monbiot, author of *Feral: Rewilding the Land, the Sea and Human Life* (2104) and advocate for rewilding, is also an admirer of what has been achieved at Knepp, but points out that there isn't enough land to make it a viable system for the UK and that producing just 54 kg of meat per hectare falls way short of what's required to feed us. He writes:

'If Knepp-style production was applied across all the UK's farmland, it would supply the UK's population with around 75 kcal per person per day – roughly 1/30th of what we need to survive.'

The question is: can rewilding projects with herbivores exist without having to kill them before the end of their natural lifespan? Obviously, without apex predators such as wolves, bears and lynx, anyone wanting to perpetuate the exploitation of animals will assume that humans should adopt this role to save the world from being overrun with wild herbivores.[45] Reinstating the predator–prey relationship to control herbivore numbers is an interesting concept, but it relies on large areas of reforestation and is therefore unlikely to happen anytime soon south of the Scottish border. Is it not reasonable, therefore, to think that populations of herding animals in rewilding systems could be controlled with contraception and euthanized when ailing after leading a full life? The argument against this would be led by a selfish desire to consume meat. Another argument against it is an economic one: since euthanized animals would not be fit for human consumpton, who would pay for such a system? From the benefits to the environment, our health and the health and well-being of the animals, it seems clear to me that we, as taxpayers, should pay for it. With much of our taxes currently being spent on farming subsidies and health care, it's not unreasonable to think that money saved in these areas could be reallocated to fund rewilding projects and animal sanctuaries throughout the country, which would in turn reinstate many of the forests and wildscapes we have lost since the Middle Ages.

Some may argue that it's an unnatural form of stewardship, but surely it is no more unnatural than intensive farm systems where battery cages, rape-racks and farrowing crates make animal lives a misery to satisfy the demand for affordable meat and dairy.[46] Biodiversity and landscape would be enriched to such a degree that it could boost eco-tourism within our own borders, which in turn would reduce the demand for holidays abroad. It might also reignite a connection with the landscape and perhaps even be the catalyst for a renaissance in home-grown food production that would not only improve well-being, but also contribute to local knowledge that might help to avert a potential food crisis in the future should fuel shortages or other issues with global trade routes make locally grown products become more imperative.

These are a few possibilities, but the potential opportunities for ecosystem restoration and rewilding allowed by transforming current food production to plant-based are vast and need to be explored in a comprehensive way.

OCEANS

While we never journeyed abroad much as a family (twice during my childhood), in 1969 we joined friends on a road trip to south-eastern Spain. The most memorable experience for me wasn't the local bar where we watched Neil Armstrong take his 'giant leap for mankind', but a local harbour where the water was teeming with fish. Shoals, like giant shadows, kept me and my brother transfixed as they ghosted their way between boats trying to evade some of the larger predator fish darting in for an easy lunch. It was mesmerizing. Later, on an isolated beach close to where we were staying, we donned masks and snorkels for the first time and entered a whole new world of fascination. While it was nowhere near as busy as the harbour, we were able to swim alongside fish within touching distance just a few metres from the beach.

Forty-plus years later, on a plane to Kalamata in the Peloponnese, that memory was uppermost in my mind. With the sea just walking distance from where we'd rented a small apartment, I could barely contain my excitement as I gathered together my aqua clobber. Doing my best to ignore the plague of plastic on the beach and a rather intimidating topless woman who strode purposefully towards us in a sort of territorial stand-off, I spat into my mask, put on my flippers and set out to explore the fruits of Navarino Bay. As it turned out, the couple of tiddlers nipping at my ankles while tightening the straps of my flippers were the highlight of the day. An hour floating about like a moody manatee revealed a lifeless bay. I wasn't prompted to learn about ocean depletion until going vegan two years later, but the feeling of unease was real. It came as no surprise to learn that between 1889 and 2007 we had already lost a staggering 94 per cent of fish stocks in the North Sea. Latest predictions are that we will have fishless oceans by 2048.[47]

Left to their own devices, our oceans are extremely efficient carbon sinks. Coastal vegetation such as seagrass, salt marshes and mangroves can capture and store carbon at a rate forty times faster than rainforests and contribute 50 per cent of the carbon buried in ocean sediments.[48] However, they are some of the most threatened natural systems in the world, with some 35 per cent of the world's mangrove forests already

At the current rate of marine destruction we could see fishless oceans by 2048.

gone.[49] A major part of this problem is clearing of mangroves for shrimp farms and, more generally, the decimation of predator populations by humans that allow unchecked herbivorous fish to decimate coastal vegetation.[50] Animal agriculture also plays a large part in this degradation as nitrogen and phosphorus from fertilizer find their way into waterways and eventually the sea, stimulating the growth of too much algae and sucking oxygen from the water as it decomposes to create ocean dead zones.[51]

The numbers of marine animals killed each year are in the trillions (up to 2.7 trillion with 55 billion fish killed by recreational fishing alone) and usually measured in tonnes. Much is discarded, as bycatch – unselective fishing methods that catch undesired animals as well as the desired species – is widespread and accidentally kills 300,000 small whales and dolphins each year.[52]

The impact of farmed fish, the largest group of farmed animals in the world, is also worrying. Fifty per cent of all fish consumed is from intensive fish farms where highly dense living conditions create the perfect breeding ground for bacteria, viruses and parasites.[53] I used to buy farmed fish thinking that it would relieve pressure on wild stocks, until I found out that farmed fish were being fed with wild prey fish and these fish are also fed not to humans but to pigs and chickens on factory farms.

With the high levels of mercury found in fish not to mention the plastic that they consume, the health implications of eating fish are also sobering.[54] It's now thought that microparticles from plastic-ingesting fish can actually get into the bloodstream of humans and that BPA (Bisphenol A – one of the chemical additives originating from plastic and a known endocrine/hormone disruptor) has the potential to cause various health issues in humans.

The recent uproar about how plastic is destroying our oceans proves that people have the capacity for caring and taking responsibility for their actions. David Attenborough's call to clean up our oceans on the BBC's *Blue Planet* series caught the imagination of people all over the world, many deciding on the spot to give up using plastic straws.[55] It sounded wonderful – people making a connection and committing to action. Consider this though: straws account for 0.03 per cent of the 8 per cent of sea pollution caused by plastics. Twenty per cent of sea pollution is caused by debris from the fishing industry like ghostnets (discarded fishing nets cause the death of 100,000 marine animals each year), ropes, traps, crates and baskets.[56] It doesn't take a scientist to work out that if you really want to save our oceans, stop eating fish. It's that simple.

WHAT CAN WE DO?

As gardeners we are told that we can make a difference. We can use less hard landscaping and give more space for plants that can store carbon. We can grow our own food and give up using chemicals. We can use plants that attract insects. We can make a pond, build compost heaps, create habitats, leave spaces to grow wild and plant more trees. We can do all these things, have fun, enjoy the view and see almost instant results. I know because I do it myself. But how do these actions alleviate some of the serious issues outlined in this chapter? What about the amount of water we use? How much deforestation and ocean depletion do we cause? How many children will go hungry and how many other species become extinct on account of our lifestyle choices? It may be an inconvenient truth for some, but our dietary choices have a profound effect on all these things and the biggest culprits are meat and dairy.

Waking up to the fact that animal agriculture is the root cause of many environmental problems can be a sobering moment. Accepting that what you have always believed to be natural, normal and necessary is, in fact, the complete opposite is challenging to say the least. A natural reaction might be to make light of it, ignore it or simply pretend that it's not your problem. But the truth is that if we carry on doing the things we've been doing for generations, the outlook is bleak for us and the generations to come.

From an individual point of view, reducing our dependency on fossil fuels and changing to a plant-based diet may not seem like a remedy for all these problems because we can't see instant results. However, most people understand that if we drive less there will be fewer vehicle emissions. One person driving less may not make a huge difference, but collectively it will. Similarly, one person giving up flying won't reduce the demand for air travel, but if we all cut down or give it up completely the laws of supply and demand mean that the number of scheduled flights would eventually fall. It follows then that one person adopting a plant based diet won't be enough to avert climate breakdown, but collectively we can reduce (and one day eliminate) the need for animal agriculture. This will have a significantly positive effect on the health of the planet and not only from the perspective of climate change.

Statistics vary and are difficult to verify, but some popular figures often quoted suggest that a meat eater will use three times as much water as a vegan, produce 2.5 times the amount of greenhouse gases and need eighteen times more land, whereas each day on a vegan diet can save 1,100 gallons of water, 20 kilogrammes of grain, 2.78 square metres of forest, 9 kilogrammes of carbon dioxide and one animal's life.[57] Perhaps more accurate is a study from Oxford researchers who found that vegans have around 2.5 times fewer greenhouse gas emissions (carbon dioxide and non-carbon dioxide) from their diets compared to meat eaters.[58] A useful way of understanding the impact of meat consumption is how it relates to car miles: how many car miles are equivalent to eating a kilo of beef.[59]

If gardeners really do care about our Garden of Eden, it's not unreasonable to assume that we have the capacity and a collective responsibility to do what we can to maintain, improve and safeguard our environment for the future. Making the switch to a plant-based diet is not only the easiest way of making a difference, it's also the most enjoyable. The question is not whether it's possible but whether we can be bothered.

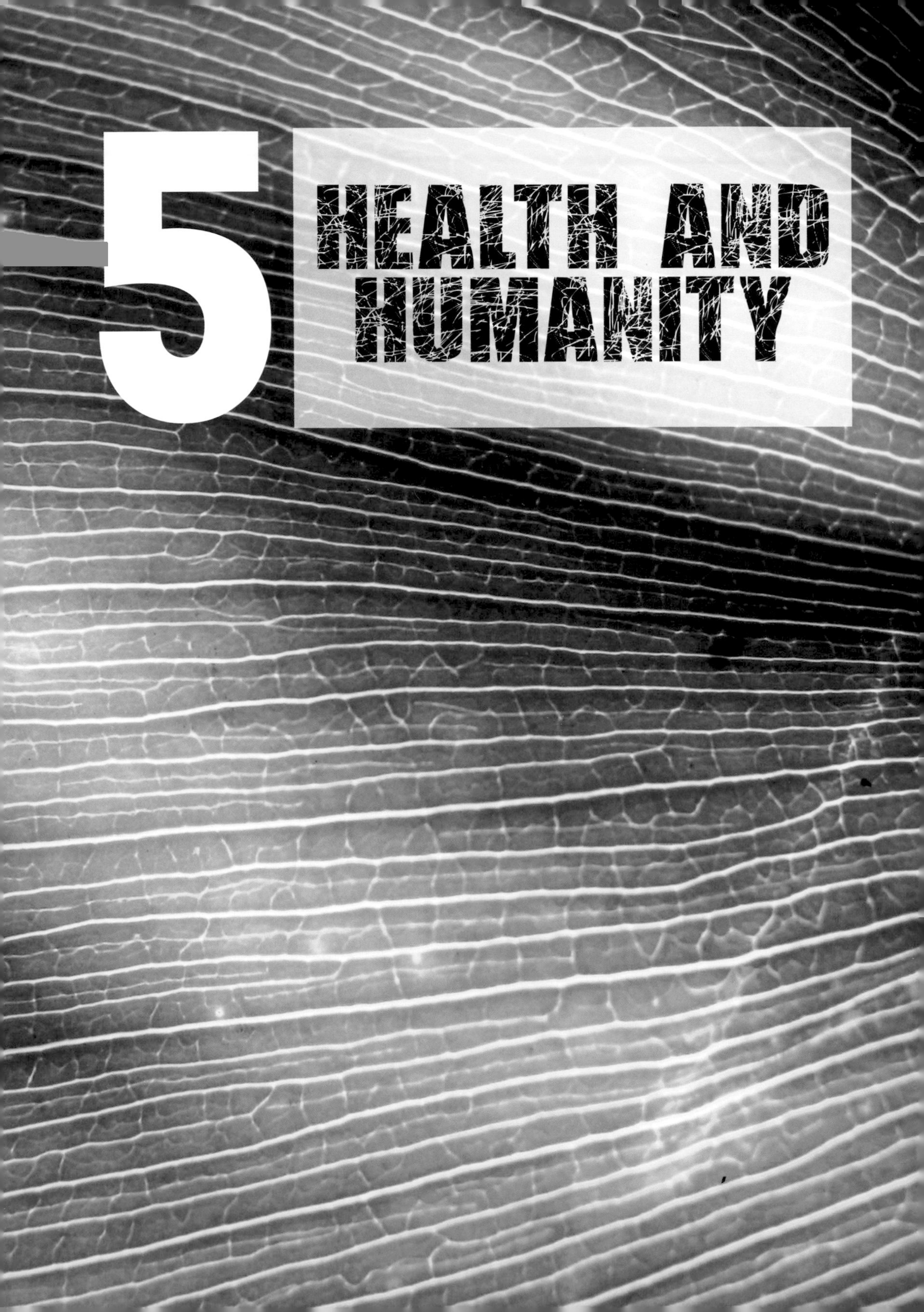

5 HEALTH AND HUMANITY

‘THERE IS MORE THAN ENOUGH PROTEIN, AS WELL AS VITAMINS AND MINERALS, IN A BALANCED DIET OF VEGETABLES, FRUIT, GRAINS, PULSES AND NUTS.’

Killed at Newman's Abattoir on 29 May 2019

While the mission for ethical vegans is to end the persecution and exploitation of animals, knowing the negative effects of animal agriculture on our health and the wider environment can encourage others to adopt a plant-based diet. It doesn't always make it easier, but whether we're indifferent to the suffering of animals or not, indifference towards our health and the health of our families – not to mention the health of the planet on which we rely to keep us alive – makes little sense.

Despite an expanding body of evidence suggesting that a plant-based diet can help prevent, treat and even reverse heart disease and other illnesses such as type 2 diabetes and high blood pressure, people are naturally cautious when that evidence flies in the face of everything they've been told. Of course that's understandable so I thought it might be worth addressing some of the concerns people might have when considering a vegan diet and highlight some associated issues.

WHERE DO YOU GET YOUR PROTEIN?

This is, without a doubt, the question more than any other that vegans regularly get asked. To be fair, I did have a few nutritional concerns of my own when I became vegan and, yes, protein was one of them. Vitamin B12 was another and then there were omega 3 fats which I'd never really paid much attention to. You see, like everyone else, I'd been led to believe that the most vital source of protein, vitamins and minerals came from meat, dairy, fish and eggs. Part of my degree was spent researching nutrition, and I don't recall any references to plant protein during my years as a track and field athlete. I've since spoken to several doctors, each of whom confirmed that nutrition isn't taught in depth during their seven years' training: about twenty-four hours in total at the most. The stark and slightly alarming fact is that a vegan who has taken time to do a little research on diet may well know more about nutrition than the average GP. (Among many others, a GP, a haematologist, a nutritionist and a surgeon have asked me where I get my protein from.) It highlights the power and influence that the meat, dairy and pharmaceutical industries have over consumers and the health service, and with successive governments continuing to subsidize these industries, profit is clearly the driving motive.

Just a little research showed that my initial concerns were unfounded. I found that there is more than enough protein, as well as vitamins and minerals, in a balanced diet of vegetables, fruit, grains, pulses and nuts.[1] Both the American and British Dietetic Associations state that a vegan diet is perfectly adequate, healthy and safe for all stages of life, including pregnancy.[2]

'LET FOOD BE THY MEDICINE AND MEDICINE BE THY FOOD.'

HIPPOCRATES

VITAMIN B12

Due to our sanitized lifestyles, B12 is the only vitamin that can't be reliably supplied by a wholefood plant-based diet. Synthesized by microorganisms (bacteria and archaea) found in the soil and natural water courses, B12 accumulates in the tissues of herbivorous animals that ingest these microbes. To a much lesser degree it is found on some plants and mushrooms. So, if you wash the mud off your vegetables, drink from a tap instead of a stream and don't consume animals or fortified products such as cereals, plant milks and some yeast products, B12 is definitely something you need to supplement.

Eating meat doesn't guarantee your recommended daily allowance of B12, however. The Framingham Offspring Study (2000) showed that 39 per cent of the general population may be B12 deficient, clearly suggesting that it isn't only a vegan issue.[3] In fact, one in six meat-eaters were found to be deficient, and those with the highest levels of B12 weren't the ones eating the most meat, but those taking supplements and eating fortified food. How can this be? Well, while meat has traditionally been thought of as a good source of B12, modern farming practices often stop the animals from producing it naturally. In fact, B12 is often given to livestock (through feed or injection) either because they have no access to grass or because of a cobalt (an essential trace element found at the centre of vitamin B12) deficiency in the soil. Heavy use of antibiotics in farm animals can also kill the bacteria that produce B12 in their gut. The most reliable way, therefore, for both vegans and non-vegans to ensure an adequate B12 intake is to take a supplement.[4]

OMEGA 3

Omega 3 is a fatty acid with anti-inflammatory effects and necessary for brain development, our immune system, nerves and eyes. Alpha-linolenic acid (ALA) is an essential omega 3 fatty acid, meaning that it can only be obtained from food. The body uses ALA to make long-chain fatty acids, eicosapentaenoic acid (EPA) and docosahexaenoic acid (DHA), which are needed for the aforementioned requirements.

A balanced and varied diet of whole foods should provide enough ALA, which is notably found in walnuts, seeds (hemp, flax, chia), oils (soyabean, perilla, echium, mustard), purslane and beans. Long-chain omega 3s are commonly found in fish and animals from free-range, non-intensive farms. The problem, however, is that mercury levels in fish – not to mention persistent organic pollutants, polychlorinated biphenyls (PCBs),[5] microplastics, micropollutants and antibiotics (in farmed fish) – negate the positive effects of DHA by impeding brain development,[6] and intensively raised animals put on weight so quickly that they can't synthesize DHA in their muscle.[7] With most meat around the world coming from animals that have been intensively raised, omnivores are just as likely to be deficient in long-chain omega 3s as vegans who don't eat a balanced diet.

And here's another conundrum. In 1976, the Royal College of Physicians and the British Cardiac Society recommended that we should reduce our consumption of red meat in favour of (leaner) chicken. This is something I hear from my more thoughtful friends who are making an effort to cut cruelty from their diets: 'I'm not eating red meat any more, just fish and chicken.' The problem here, apart from the obvious one – that chickens and fish have just as much capacity to suffer as pigs, cows and sheep – is that the genetic selection and breeding of birds (not to mention growth hormones and a lack of exercise) mean that there is a disproportionate amount of fat in chickens today – three times more fat than protein.[8] Furthermore, aside from the increased cancer risks associated with chicken,[9] the amount of DHA from a 100 g serving of chicken has dropped from 170 mcg in 1980 to 25 mcg in 2004.[10]

A reliable source of long-chain omega 3s for both vegans and omnivores (who eat intensively raised animals) is where the fish get it – algae. A pollutant-free supplement made from algae means you don't have to eat fish.[11] Also, aside from all the environmental damage associated with fishing, not ingesting the heavy metals and other pollutants found in fish will also improve your health. Giving up eating fish is nothing less than a win–win situation on many levels. Having said all that, the most recent data available at the time of writing suggests that as long as vegans eat a balanced, wholefood, plant-based diet there's no need to supplement with omega 3, and that the science behind any advice to supplement is lacking.[12]

WHAT ABOUT THE OTHER PILLS?

So, to be clear, if you're vegan eating a balanced, wholefood, plant-based diet, the only supplement you might need is B12. While I'm reasonably confident that I get enough B12 from cereals, plant milks and nutritional yeast, I regularly take a supplement to be on the safe side.[13] But the burning question is this. If the only supplement vegans are being advised to take is B12, who is taking all the other vitamins, minerals and protein drinks that we see on the shelves? Supplements can be expensive so it's safe to assume that the people buying them can afford to eat reasonably well. Supermarkets, chemists and health stores have whole aisles dedicated to vitamins, minerals, amino acids and other dietary supplements, the large majority of their customer base being made up of omnivores and vegetarians eating the supposedly nutrient-dense meat and dairy. Are these foods lacking in the vitamins they are supposed to provide, or does a cocktail of vitamins alleviate the discomfort, reaction or illnesses they cause? Of course, it could simply be clever marketing by the health industry to make consumers take supplements unnecessarily. All these companies have to do is sow the tiniest seed of doubt that people are in danger of being malnourished if they don't take vitamin X, Y and Z and you'll have their attention. It is a ploy that the meat and dairy industries use all the time to make people believe

'ONLY ONE WAY OF EATING HAS EVER BEEN PROVEN TO REVERSE HEART DISEASE IN THE MAJORITY OF PATIENTS: A DIET CENTERED AROUND WHOLE PLANT FOODS. IF THAT'S ALL A WHOLEFOOD, PLANT-BASED DIET COULD DO – REVERSE OUR NUMBER-ONE KILLER – SHOULDN'T THAT BE THE DEFAULT DIET UNTIL PROVEN OTHERWISE? THE FACT IT MAY ALSO BE EFFECTIVE IN PREVENTING, TREATING, AND ARRESTING OTHER LEADING KILLERS SEEMS TO MAKE THE CASE FOR PLANT-BASED EATING SIMPLY OVERWHELMING.'

DR MICHAEL GREGER, 'Plant-based diets', nutritionfacts.org

that they need their products for optimum health. The difference between the two scenarios is that the unnecessary consumption of meat and dairy products comes with risks to personal health, damaging consequences for the environment and unavoidable suffering for animals, whereas the unnecessary consumption of supplements may just give you expensive, day-glow pee.

NUTRITION FACTS

So why aren't we told more facts about nutrition? Why aren't we taught about preventative medicine through diet? Why don't doctors tell us to avoid saturated fat predominanatly found in animal products and that we should instead choose a plant-based diet to maintain optimum health? The main reason is that doctors aren't taught about nutrition in any great detail at medical school. With the pressures the NHS is currently facing and the information we now have at our fingertips, this is nothing short of scandalous.

However, one only has to look at how long it took for people to believe that smoking is bad for you to see the bigger picture. The meat and dairy industries are big business and, unlike broccoli and leafy greens, make money. Lots of it.[14] The pharmaceutical industry also profits massively, not just from drugs that are used to try and cure us, but from drugs that merely relieve the symptoms and keep us ticking over so we can consume more of what is making us sick in the first place. Add to that the 450 drugs administered to the animals themselves[15] and you suddenly have the perfect self-sustaining business model, which is why anyone with a financial interest in such a market will find the very notion of veganism a threat.

Perhaps even more disturbing is the idea that people will ignore this information at the expense not just of their own health but of that of their families, the seriousness of which is made clear by Dr Michael Greger, founder of Nutritionfacts[16] and author of *How Not to Die*: 'By age 10, nearly all kids have fatty streaks in their arteries. This is the first sign of atherosclerosis, the leading cause of death in the United States. So, the question for most of us is not whether we should eat healthy to prevent heart disease, but whether we want to reverse the heart disease we may already have.'[17]

DAIRY

Perhaps the biggest revelation to me in becoming vegan was just how unhealthy dairy products are. From childhood we are told that cow's milk will give you strong bones, but scientific data doesn't support this and some studies suggest that it increases the risk of mortality and fractures.[18]

While milk, cheese, butter, yogurt and ice cream are a source of calcium, this one positive is overwhelmingly outweighed by the negative baggage such as cholesterol, saturated fat, antibiotics and growth hormones.[19] In the film documentary 'What the Health', Michael Klaper MD, who was raised on a dairy farm, explains that the purpose of cow's milk

> is to turn a 65 lb calf into a 400 lb cow as rapidly as possible. Cow's milk is baby calf growth food, that's what the stuff is! Everything in that white liquid, the hormones, the lipids, the sodium, the protein, the growth factors like IGF-1, every one of those is meant to blow that calf up into a great big cow - or else they would not be there. And whether you pour it on to your cereal as a liquid, churn it into butter, curdle it into yoghurt, ferment it into cheese or freeze it into ice cream, it's baby calf growth food! Its purpose is to increase weight and promote growth in tissues throughout the mammalian body. It's great stuff if you are a baby calf, but if you are a human trying to create a lean, healthy body, it will NOT do a body good.

Dr Klaper goes on to explain how cow's milk stimulates the tissues in women giving them breast lumps, an enlarged uterus, bleeding and fibroids, which can lead to hysterectomies. Milk products have been linked with man-boobs in men, and studies have suggested links with a number of allergies and even autism and cot death.[20]

There's also the issue of pus. With the continuous cycle of artificial inseminations and mechanized milking (for an average of three years before they are killed for meat), it's hardly surprising that cows are often affected by mastitis, an inflammation of the udder caused by bacterial infection.[21] In the UK, according to Compassion in World Farming, there may be up to seventy cases of mastitis each year in a herd of one hundred cows. In the US mastitis epidemics are common, with one in six cows affected. Around 90 per cent of somatic cells in an infected cow's milk are neutrophils (inflammatory immune cells) that we know as pus.

The good news is that there's not nearly as much pus in milk as some vegans would have you believe. You see, when I first started looking into this, there were reports of a vial full of pus in each glass of milk. It turns out that it's no more than a drop of pus per glass.[22] The bad news, however, is that it's still pus, and there's more than a drop of it in more concentrated products like cheese, butter and cream. Now we all know that cow's milk is pasteurized, so the pus and any faecal matter (sorry, I forgot to mention the poo) that may have found its way on to the cow's teat is sterilized, but cooked, curdled, shaken or churned, pus is pus, poo is poo, and I'm not inclined to want either in my beverage or food no matter how much it's been heated.

Anyway, enough about pus and poo. Let's talk about something slightly more uncomfortable: breast-feeding. The one huge revelation for most vegans is that the only milk we've ever needed is from our mother. It's common sense, pure and simple, but to anyone as old as me who can remember getting free milk at school, I know this must sound completely bonkers. As children, we were told over and over again by people we trusted that cows' milk is necessary for healthy development. The truth is that once weaned from our mother's milk, we have no further use for it. And, let's be honest with ourselves here. I mean *really* honest. Isn't the idea that we should continue to breast-feed throughout our adult lives more than a little weird? Isn't it disturbing or, at the very least embarrassing that we should seek secretions from the teats of another species?[23] Doesn't the fact that we pay people to spend a lot of time fiddling with a cow's private parts in order to take milk meant for her baby seem a little perverted? How long can something that is, let's face it, nothing short of deviant be dressed up as being normal and acceptable?[24] With the misery that cows have to endure for us to take their milk,[25] the aforementioned health consequences from consuming it, and the greenhouse gas emissions and general environmental issues associated with the rearing of cattle, you have three perfectly good reasons to wean yourself off the habit.

There are signs that the tide is turning. The dairy industry is on the wane, in developed countries at least. Between 2013 and 2016 over one thousand dairy farms in England and Wales closed down, and between November 2018 and January 2019 a further 116 producers departed.[26] In January 2019, Canada omitted dairy from its food guide. After three years of consultations and ignoring reports funded by the dairy industry, its evidence-based conclusions encouraged instead a variety of unprocessed foods. Canadians are now advised 'to fill half their plate with fruit and vegetables, a quarter with starches or grains and a quarter with protein'.[27] While this is a step in the right direction,

Cows naturally mourn for days after their calves are taken.

from the cow's perspective it's not all good news, as industrialized farms and increased yields per cow will make up the shortfall. Heavily subsidized, the dairy industry can no doubt survive competition from plant-based milks but for how long? Profits are at stake, and pushback from the dairy industry is inevitable with misleading claims about the health benefits of milk and related products. However, the news that Dean Foods, the USA's biggest milk processor, filed for bankruptcy just as we put this book to bed means there's reason to be optimistic that dairy, perhaps the most exploitative and cruel of all industries, will one day be consigned to the history books.[28]

PHYSIOLOGY

In October 2015 a World Health Organization (WHO) report stated that processed red meat has been classed a grade one carcinogen (the same group that includes tobacco, alcohol and asbestos) usually associated with colorectal, pancreatic and prostate cancers.[29] Red meat on its own is classed as a grade two carcinogen. While meat is nutrient dense it has no fibre and comes with unwelcome baggage (cholesterol, saturated fat) which can also cause heart disease and type 2 diabetes. Aside from these findings there is good evidence that we're not designed to eat meat. Our digestive tracts are much longer than those of carnivores and omnivores[30] and our jaw structure is much weaker; our canines may cope with fruit and nuts, but they are hopeless when it comes to ripping raw flesh, tendons and bone like an omnivore, and while the sight and smell of flesh and blood would make an omnivore salivate, it's offensive to our senses and makes us retch.[31] Dr Milton Mills suggests that this natural feeling of revulsion when seeing flesh is there to protect us from sickness, and explains that humans have the anatomy and physiology of a strict plant-eater or herbivore. He notes that our observation of what people choose to eat leads us to the incorrect assumption that we are natural omnivores: 'We don't have any adaptations in our digestive system or our physiology that is adapted to eating or consuming animal flesh. And that's why we can't consume animal flesh without the aid of technology.'[32]

DIABETES

Over the last few of my advancing years I've heard many people talking about their ailments, the most common being type 2 diabetes. It's hardly surprising. Over two million people in the UK suffer from this disease, which can cause blindness, kidney failure, heart attack or stroke. In the USA, 100 million are now living with diabetes or prediabetes.[33] Today, with developing countries emulating western diets, type 2 diabetes is nothing less than a global epidemic which is pushing some health care systems to breaking point.

Type 2 diabetes is often linked with obesity and a sedentary lifestyle, which means it's nearly always preventable and can be treated, and sometimes even reversed, through diet and exercise. Until fairly recently, the assumption has been that diabetes is caused by sugar or high-carbohydrate diets and that cutting down on bread, rice and other starches (that release glucose into the blood) is the most effective way of controlling the disease. Research by Dr Neal Barnard and his team at Physicians Committee for Responsible Medicine, however, has shown that animal fat is the more likely cause as it builds up in your muscle cells and in your liver, making it more difficult for those cells to absorb glucose.[34] Insulin, a hormone that attaches to receptors on muscle cells, acts like a key to unlock the receptors and allow the sugar to enter the cell. Too much fat in the cell and the insulin sensitivity of these receptors is inhibited, preventing sugar from getting through which, in turn, causes problems by raising blood sugar levels. Their research is backed up by historical evidence, most notably in Japan where, before 1980, there was a very low incidence of type 2 diabetes, with just 1–4 per cent of the population suffering from the disease. By 1990, ten years after McDonald's set up shop in Japan, diabetes skyrocketed to 11–12 per cent of the population.[35]

Those on a plant-based diet are less likely to suffer type 2 diabetes than people who regularly consume meat.[36] This is largely due to the fact that they tend not to be obese and don't consume saturated fats that contribute to insulin resistance. They also consume more monosaturated fats (found in nuts, seeds and avocados) that are thought to protect against the negative effects of saturated fats. Vegans, therefore, seem to have better insulin sensitivity, lower blood sugar levels and higher insulin levels.[37]

While Barnard's research and advocacy for preventative medicine through a plant-based diet have helped many of his patients with type 2 diabetes (some making full recoveries with no need for further medication), he's careful to point out that vegans also have to watch their diet and not overdo it with oils, salad dressings and frying oils which can also clog cells and inhibit insulin receptors. Refined starches (which have less fibre) should also be consumed in moderation, so you might want to substitute sweet potatoes for potatoes, brown rice for white rice, wholemeal flour for plain flour. Barnard's research has shown that a well-balanced, plant-based diet can also help conditions such as heart disease, stroke, arthritis and Alzheimer's disease.

Note: if you're on medication for diabetes, you must tell your doctor that you are switching to a plant-based diet so that prescription drugs can be monitored and reduced as things improve. If you don't, there's a risk that your blood sugar levels could drop too low.

ANTIBIOTICS

As a sluggish writer and expert procrastinator, deadlines bother me.[38] So, just before the Easter break of 2019 I decided to clear the decks and take advantage of the holiday period to get to a place that felt just a little bit more comfortable. After just one day I was hijacked by an infection. Fever, night sweats, disorientation and a headache from hell had me wondering if an unknown tropical stowaway, which had put me in and out of hospital over a course of five weeks, had resurfaced after spending twelve years dormant in my liver.

Two weeks later, when I was suffering mild hallucinations and was unable to function, an overdue trip to the GP brought good news. It wasn't the alien sleeper I'd feared, after all; it was pneumonia. I'd never taken antibiotics before, but with mixed feelings of reluctance and relief I gratefully accepted the prescription. My GP warned me that I would feel a whole lot worse before getting better (he was right about that), but I had no choice and was incredibly thankful for this drug that I'd spent most of my life trying to avoid. As the fug began to clear, I thought how much we take antibiotics for granted and about the feared scenario where our overuse of them could potentially cause a global health crisis as bacteria become more resistant. The problem is further exacerbated by the fact that animals in intensive systems are routinely given antibiotics. In fact, 70 per cent of antibiotics in the USA (medically important for humans) are used for animals.[39] And, with increasing wealth and demand for meat in emerging countries, the problem is likely to get even worse.

Antibiotics are not only given to sick animals; they are routinely given to healthy ones to prevent infection in the overcrowded conditions of intensive farming. They are also used to speed up growth to make meat more affordable. Around half of all antibiotics used globally are given to farm animals.[40] A UK review of antimicrobial resistance (AMR) found that out of 139 peer-reviewed academic papers, 72 per cent found evidence of links between antibiotic consumption by animals and resistance in humans and that this justified a call for a global reduction in the use of antibiotics in animals.[41] It can't happen soon enough. In the UK 5,000 people die every year from antibiotic resistant E. coli, and indications are that the use of antibiotics on farms is one of the main causes. There are also concerns about drug-resistant strains being passed to humans when they prepare or eat meat as well as environmental pollution from antimicrobials excreted by the animals and waste from the manufacture of the drugs being discharged into water courses.

HUMANITY

While the effect of animal agriculture on the environment and our health is well documented, the effect on humanitarian issues is not so obvious. Animal rights activists are often confronted by people saying that human rights are more important. Of course, ignoring the fact that humans *are* animals, it's perfectly possible one can campaign for equal rights for all beings. It's worth considering, though, that human rights are being affected every day by animal agriculture. For example, 82 per cent of starving children live in countries where food is fed to animals, and the animals are eaten by western countries. With so many people going hungry, is it not criminal that 50 per cent of the world's grain is fed to animals? If this food was fed directly to humans, then the predicted rise in world population (9.8 billion by 2050 and 11.1 billion by 2100) would not seem quite as daunting.[42]

The devastating effect of animal agriculture on climate change is already having severe consequences for people living on islands and coastlines where sea levels are rising and in countries where there are water shortages. Forest fires and megastorms fueled by increased temperatures have claimed many lives and homes. Pollution from pig farms where lagoons of effluent are sprayed across fields near low-income families causes acute respiratory diseases, and the sufferers have no means of recourse.[43]

Then there are the slaughterhouse workers themselves. Paid at the minimum wage and under pressure to kill hundreds or even thousands of animals a day, it's hardly surprising that some of the highest rates of work-related injuries and fatalities occur in slaughterhouses, where the pressures to keep the production line working efficiently make the job of killing extremely hazardous. They also have some of the highest rates of domestic violence, drug and alcohol abuse, PTSD, anxiety, depression and suicide in any industry.[44] A study in the USA of the effect of industries on communities found that those with slaughterhouses had a marked increase in overall crime with disproportionate increases in violent and sexual crime.[45]

Animals sent to slaughter have only lived a small fraction of their natural lives. They are literally babies. These beautiful piglets were killed at Newman's Abattoir on 13 December 2017.

Slaughterhouse work is a dirty job and no one wants to do it, so predictably much of it falls on the shoulders of migrant workers. European Union migrants account for 69 per cent of the British meat processing workforce; many of them don't speak English well enough to stand up for themselves when they are being exploited. In 2016, when there was a shortage of people willing to work in Canadian slaughterhouses, the government looked to Syrian refugees to fill the spaces on the kill floor. The cynicism in welcoming refugees fleeing the violence in their war-torn country only to offer them work killing animals all day is a sad indictment on modern society. Chas Newkey-Burden sums it up perfectly in an article for the *Guardian*:

> Most people don't like to think about the effect that buying meat has on animals and the environment. Few are even aware of the plight of slaughterhouse workers. But market forces are simple – every time you put meat in your shopping trolley, you are funding the slaughter, globally, of 70 billion farmed animals each year, the destruction of the environment and, yes, the exploitation of vulnerable workers.[46]

SHARING KNOWLEDGE

Even when facts are acknowledged, our addiction to meat and dairy is so strong that it overrides any concern for personal health or even the health of our families. A friend, on accepting the health risks of bacon and other processed meats, said that he would still feed it to his children and 'let them take their chances'. I found such an honest declaration from a seemingly responsible parent shocking, and it made me realize just what the vegan movement is up against: cognitive dissonance and a refusal to change a habit even when the health of one's own child might be at stake.

The good news is that we now have the knowledge to take control of our lives where health is concerned. Just like the tobacco industry, the dairy and meat corporations will continue to promote their agenda because they see it only from an economic point of view. However, the evidence that supports the health benefits of a plant-based diet is overwhelming and can be ignored for only so long.

The vegan movement has been labelled a fad and a trend, and any suggestion that a plant-based diet is best for your health is met with scepticism and even ridicule. This is driven not just by a lifetime of indoctrination but also by the simple fact that meat is tasty and it makes money. I've often heard people say, 'I don't really like vegetables but I eat them because they are good for me,' but I've not met anyone who says, 'I don't like meat, but I eat it because it's good for me.' The bottom line is that when we stop consuming animal products and adopt a varied plant-based diet we are healthier for it. The word 'varied' is, of course, important here.

Vegans who live mainly off processed food high in fat, salt or sugar will almost certainly be more unhealthy than those who eat a mixed wholefood diet using fresh vegetables, fruits, grains, nuts and legumes. It's also worth pointing out that while a plant-based diet can reduce your chances of getting diet-related diseases like heart disease, cancer, diabetes and high blood pressure, it doesn't necessarily reduce your chances of getting diseases caused by other factors such as pollution, stress and smoking.

Meat and dairy consumption is on the decline in developed countries, and the market for meat alternatives is expanding as people look for healthier sources of protein.[47] A shift towards a plant-based diet is already happening, and as the food industry is realizing that there is a lucrative market for these products, the supply of alternatives to meat is increasing and the demand for meat will eventually decrease. Unfortunately, this coincides with a rise in meat consumption in developing countries where new-found prosperity has increased demand for animal products, so there we will probably see an overall increase in animal consumption before a significant global reduction.

I could fill the rest of this book with independent scientific studies and peer-reviewed papers that show just how much healthier a vegan diet is for you, but from experience, facts and figures tend to be debunked by carnists quoting their own statistics, often sourced from research funded by the meat and dairy industries. Please don't take my word for it. Instead, look no further than Dr Michael Greger's book *How Not to Die* or watch some of his lectures and videos via his evidence-based, non-profit, no adverts website Nutritionfacts.org, where thousands of peer-reviewed papers make this a goldmine of nutritional information.

The only problem with this knowledge is, of course, how to share it. Even when you have the interest of the person at heart, they are immediately suspicious of your intentions. I've had several conversations with friends and relatives (some of whom have suffered ill health), and my suggestion to try a plant-based diet has fallen mostly on deaf ears, almost to a point where I've more or less given up suggesting what could be useful information on how to stay alive. It raises an interesting question. If a friend or relative was ill with heart disease or type 2 diabetes and you knew of a medicine that could not only alleviate their disease but potentially cure it and there wouldn't be any extra expense involved, would you tell them about it? I'm sure they'd like to think that you would. What sort of a friend or close relative wouldn't share something that would ultimately improve the health of a loved one? The problem is that even if they've known you their whole life and trust you implicitly, the mere suggestion of a plant-based diet is likely to raise a suspicion that you're giving them false information to trick them into becoming vegan. It's both frustrating and disturbing, and shows just how effective propaganda from these exploitative industries actually is.

NATURE MAKING US FEEL WHOLE

Have you ever noticed how less than wholesome you feel when someone (human or animal) close to you is ill? Also, when we see pollution, deforestation and development on a grand scale, doesn't part of us wither inside when we witness it? Conversely, when someone recovers from illness, the sense of relief and well-being among a family or carer is tangible. When we walk through pristine forests and swim in oceans without seeing the faintest shred of litter, the experience is elevated by the impression of purity suggesting that all is right with the world. Our health is intertwined with everything and everyone around us. When connections are whole we feel wholesome. When they are broken it wounds us. If we try to pretend that things are okay and hide the truth, we still know deep down that things aren't right. Turning a blind eye to broken connections or making light of them somehow normalizes them.

Being aware of your surroundings also affects the way we see things. You only have walk through an airport or train terminal to see the lack of spatial awareness people have when trying to get to their next destination as quickly as possible and without any care for others who might be in their way. Some people, of course, are more in tune with their surroundings than others. Gardeners especially. Gardeners are drawn to nature and their immediate environment. Many of the gardeners I've known have described how they were beguiled by an invisible force and a deep desire to be intimate with nature or at least work with it in some meaningful way. We know how gardens can make us feel whole and how even the most tenuous link with nature can lift us when we are troubled or challenged.

When someone suggested I should adopt a vegan lifestyle, I thought they were trying to take something from me. I was wrong. They were offering me a gift. And a priceless one at that. Physically I'm healthier and my fascination with the natural world has been enriched. Adopting a vegan lifestyle had an unexpected effect on my understanding and appreciation of nature. It's almost as if everything went offline for a moment and, on rebooting, I found that the signal had been enhanced to a more sensitive system. I hesitate to say it's a spiritual connection, but while there is a sense of being plugged-in, it's more like finding the last piece to a jigsaw that I had no idea I was building.

As a gardener I have always had a sense of being at one with nature, but now it's on a different level. It's an awareness of the vital connections between everything that lives and the invisible forces that are constantly at play interacting with all life forms to maintain an illusion of calm. Darryl Moore of Cityscapes alluded to such connectivity at the first Vegan Garden Festival in 2018:

> **Gardens are far from the static tranquil places we imagine them to be. A picture of serenity masks a frenetic stream of challenges faced by plants, which threaten their survival and the healthy balance of wider environments. Understanding the interconnectedness between plants and other species is essential for developing a sustainable plant-based perspective of the world beneficial to all life forms.**

Even with the best of intentions, our collective actions and interventions can wreak havoc with the very things we'd like to protect. However, with careful consideration we can effect positive changes too.[48]

We have to be open-minded and willing to change, but change, of course, is the overriding problem. Humans don't like it one little bit, especially when it's not from their choosing. But change can be good. The changes I have made have opened my mind to what we have: a beautiful, fragile world whose existence is nothing short of a miracle. Our expectations of what we want from life in terms of material wealth, careers, children and holidays have to change too. It's clear what we stand to lose if we don't act now. From the research for this book alone, I can't honestly say that I'm 100 per cent hopeful. We are in the midst of the sixth mass extinction and have the knowledge to offset it, but do we have the inclination? Clearly, there are many people in the world who have more than enough trouble making ends meet and getting through the day to worry about what the future might hold. It's the conversations I've had with people who clearly don't have these problems but are either oblivious to the effect we are having on the planet or aware but completely indifferent that are more concerning.

Gardening isn't just about pretty flower and bees. As Darryl Moore says, we are dealing with other life forms. 'They are an integral part of the success of our ideas and designs, whether they are plants, animals, invertebrates, microbes or bacteria. We need to ensure that we are creating beneficial balances and interactions with all these "stakeholders" and this entails a form of empathy based on respect and care of duty. It also entails dismantling the established hierarchies that objectify and denigrate the nonhuman life forms that we share the environment with.'

6 FOOD

‘WE CAN ACT AS A CATALYST TO BRING ABOUT AGRICULTURAL, CULTURAL AND SOCIAL CHANGE. MUCH EMPHASIS IS NOW PUT UPON THE EVIDENCE THAT SUGGESTS THAT A WHOLEFOOD, PLANT-BASED DIET IS HEALTHY. THIS DIET, BASED ON WHOLEFOOD GRAINS, TUBERS, FRUIT AND VEGETABLES, IS ALSO THE MOST AFFORDABLE.’

JENNY HALL, ‘GROWING VEGANICALLY: BECAUSE BEING VEGAN ISN’T JUST ABOUT WHAT WE EAT!’, VEGAN ORGANIC NETWORK (2019)

Killed at Newman's Abattoir on 13 December 2017

In March 2019 a vegan pie beat 886 competitors to be crowned overall champion at the British Pie Awards. What was amazing wasn't so much that it won, but the backlash from a tearful Michelin-star chef and the usual bunch of anti-vegan 'celebrities'. It was hardly surprising. Just a couple of months earlier, the anti-vegan lobby had gone into meltdown when Greggs announced the launch of its vegan sausage roll. Eight months later, the Greggs CEO, Roger Whiteside, gave up meat and dairy after seeing the film *The Game Changers* and is now ramping up efforts to offer more vegan options at the UK's most popular bakery chain.[1]

The idea that someone might get upset when no animals have to be killed for a fleeting gastronomic pleasure is a curious one. The notion that you're going to be deprived of some of the tastes and textures you enjoy by switching to a plant-based diet may have been true twenty years ago, but not any more. Vegan burgers, sausages, ribs, mince, fish and cheese mean that you really don't have to give up any of the tastes you've grown to know and love. Plant-based business is booming and it's rattled the meat and dairy industries.[2] The US state of Mississippi is trying to ban the use of the words 'Vegan Burger' and 'Vegan Hot Dog' on food labels so as not to confuse consumers. Here in the UK the dairy industry threatened action against a London vegan cheese shop, La Fauxmagerie, arguing that labels describing their product as cheese are misleading the public.[3] While this doesn't say much about the intelligence of consumers, it does suggest that the rise of veganism is beginning to hurt the larger corporations that profit from animal agriculture.

I'm often asked why a vegan would want to eat something that resembles the flesh of an animal. It's a good question but there's a simple answer. While there are some vegans who can't bring themselves to eat these substitutes, for others (who enjoyed the taste and textures of these products) it makes the transition from animal to plant-based food so much easier. Fortunately, many of these meat substitutes like burgers, balls or sausages are literally shapes that don't resemble body parts at all and can easily be made using vegetable protein instead.

'PEOPLE ARE REALLY OUT THERE WORRYING ABOUT HOW PLANTS BECOME BURGERS WITHOUT HAVING A CLUE HOW ANIMALS BECOME BURGERS.'

@herbivore_club, Twitter (17 October 2019)

Some meat substitutes, despite being healthier than the meat versions with no cholesterol or saturated fat, are still 'processed' and are likely to contain more oil and salt than you would ordinarily use when cooking at home. They can also be expensive because at the moment, unlike the meat and dairy industries, they are not subsidized. While everyone (even vegans) enjoys a bit of junk food from time to time, the benefits of a balanced vegan diet to health and well-being shouldn't be underestimated. Of course, not everyone has the time or inclination to cook for themselves, but those who do will know that a vegan diet is nourishing and inexpensive and, with a little imagination, can be quick to prepare and delicious.

'BEFORE I BECAME VEGAN MY FRIDGE WAS A MORGUE WITH DEAD CHICKEN AND PIGS . . . NOW IT'S A GARDEN WITH COLOURS AND TEXTURES AND THINGS THAT I'D NEVER BEEN AWARE OF BEFORE.'

EARTHLING ED, 'You'll never look at your life in the same way again', YouTube (2018)

A plant-based diet might sound restrictive and dull, but most vegans will tell you that it's quite the opposite as the range of food you eat will increase significantly. 'Five-a-day' becomes more like fifteen a day as you suddenly become aware of how restricted your diet was before. Many of the basic food staples are naturally vegan and cheaper than meat and dairy products. Bread, rice, potatoes, vegetables, fruit, lentils, grains and beans can be used in a wide variety of ways. Again, not everyone will be inclined or have the time to learn new recipes, but there are plenty of resources to help you prepare quick and easy meals (see the resources section). YouTube and social media sites like Instagram have opened a whole new world of cuisine with a seemingly limitless repertoire of vegan cuisine from around the globe. The following recipes are a few of the ones we go to when time is short. They're relatively simple to make, easy on the pocket, and can be adapted fairly easily to fit around what you have in the cupboard at any given moment. Don't forget that while a balanced, wholefood, plant-based diet is generally healthy, even vegans are tempted by junk food. Processed food such as plant-based burgers, fried food and cakes can have high salt, fat (vegetable oil) or sugar content and should be consumed in moderation. An unhealthy vegan isn't a good advert for the movement.

SOME USEFUL VEGAN STAPLES TO HAVE IN YOUR CUPBOARD:

- Vegetable stock – Marigold Organic Swiss Vegetable Boullion Powder: we use this for soups and a range of other dishes.
- Marigold Savoury Engevita Yeast Flakes, a cheesy/nutty-tasting product fortified with vitamin B12: we use this in soups, sauces, risottos and paella.
- Bragg's Liquid Aminos (contains sixteen amino acids): we use this mostly in stir-fries as a soy sauce alternative.
- Flax seeds: soaked seeds can be used as an egg replacement and a useful binding agent in baking.

TOFU SCRAMBLED EGGS

Serves 2–3

Muesli, fruit and occasionally toast comprise our habitual breakfast. Any desire for eggs disappeared completely once I understood the full picture behind the egg industry, but this is a quick and easy vegan version of scrambled eggs which we make from time to time. Not only is it cruelty and cholesterol free, it's tastier too.

Ingredients

1 tsp coconut oil
4 spring onions, finely chopped
½ tsp turmeric
6 cherry tomatoes, sliced into quarters
1 tsp Marigold stock mixed with 100 ml warm water
1 tsp Marigold Engevita Yeast Flakes
1 300 g packet of silken tofu
Pepper
Chopped coriander or parsley to taste
Toast

Method

Heat oil in a non-stick frying pan. Fry onions for a few minutes, then add turmeric. Let the turmeric cook for a minute to colour the onions, then add the tomatoes.

Cook so that the tomatoes break down, then add Marigold stock. Stir to mix the juices, then add the silken tofu, gently breaking up the tofu as you stir. The tofu will take on the colour of the turmeric. Be careful not to break the tofu too small. Cover and simmer for five minutes on a very low heat.

Switch off heat, add chopped coriander or parsley, pepper to taste, and leave to rest for a couple of minutes before serving on toast with an extra sprig to garnish.

PANCAKES

Makes about 6 medium pancakes

We used to make pancakes as a treat from time to time. Christine prefers a thin, crepe-like pancake while I like the thicker American ones. Finding the right recipe took a bit of time but it was worth the wait. Thank you, Emma Doherty, for this one.

Ingredients

145 g wholemeal flour
1 tbsp baking powder
230 ml oat milk
2 tbsp coconut oil plus a little extra for frying
2 tbsp maple syrup
1 tsp apple cider vinegar
1 tsp vanilla extract

Method

Mix dry ingredients in one bowl. Mix wet ingredients in another bowl. Pour wet into dry but don't overmix. Let it sit for 5 minutes and then cook pancakes in a lightly oiled non-stick pan. After the first pancake you will need even less or preferably no oil.

Stack with blueberries, bananas, soy or cashew cream or whatever you fancy with more maple syrup to taste.

SPICY TOMATO SOUP

Serves 6

This is a slightly adapted version of a recipe by Julia Sahni in her amazing book *Classic Indian Vegetarian and Grain Cooking*.

Pre-cook 180 g of red lentils in 750 ml of water. Add ½ tsp of turmeric and bring to a boil, stirring occasionally. Partially cover pan and boil on a medium heat for 25 minutes. Reduce to a low heat, cover and cook for a further 10 minutes.

Ingredients

500 g tinned or ripe plum tomatoes
The cooked red lentils
250 ml water
1½ tsp ground cumin (grind fresh seeds)
2 tsp ground coriander (grind fresh seeds)
½ tsp cayenne pepper
1 tbsp minced onion
1 tsp minced garlic
1 tsp Marigold boullion
6–10 cherry tomatoes, halved
1 tbsp lemon juice
Spice liquor ingredients:
1 tbsp coconut oil
1 tsp black mustard seeds
10 fresh curry leaves (or dry)

Method

Drop fresh tomatoes (if using) in boiling water for two minutes or until their skins burst; remove skins and blend in a liquidizer or simply blend tinned tomatoes if using.

Combine blended tomatoes and cooked lentils in a large pot and stir. Add all ingredients except for lemon juice, tomato halves and spice liquor and bring to a boil. Lower heat and simmer for 10 minutes. Add lemon juice and tomato halves, and cook for a minute or two longer before covering and turning off the heat.

For the spice liquor, heat the oil until hot, then fry mustard seeds and curry leaves. The seeds will pop and the leaves spatter so keep a lid to hand just in case. Once it's calmed down, pour the liquor into the tomato soup and stir.

If the soup is too thick add a little more water.

TOMATO SAUCE

Serves 2

When time is short there are a whole range of tomato sauce products you can go to. However, when there is time to make your own, you can have better control over the amount of salt that's added.

Ingredients

1 tbsp olive oil
1 onion, sliced and finely chopped
4 cloves garlic, finely chopped
1 carrot, grated
1 red pepper, roasted, burnt skin removed and chopped
1 stick of celery, finely chopped
1 tsp Marigold boullion
A squeeze of tomato concentrate
200 ml red wine
6 fresh medium-sized tomatoes or a 400 g tin
200 ml water
Pepper and herbs to taste

Method

Heat oil in a pan. Add chopped onion and cook until translucent. Add garlic and fry for a minute. Add carrot, red pepper and celery. Mix well and fry for a few more minutes.

Add Marigold boullion and tomato puree and mix well before adding the red wine. Cook for a couple of minutes then add the tomatoes and water.

Season with pepper and a pinch of dried herbs (such as basil and marjoram), cover and simmer for 30 minutes, stirring every so often and adding a little water if too thick.

CHICKPEA CURRY Recipe courtesy of Samantha Meah

Serves 4

Ingredients

2 tsp cumin seeds
2 tsp coriander seeds
2 medium onions, chopped
5 garlic cloves
2.5 cm piece of peeled fresh ginger, grated
1 tsp turmeric
1 tsp chilli powder
2 400g tins of chickpeas
4–5 fresh chopped tomatoes
500 ml warm water mixed with 2 tsp Marigold bouillion
A bunch of coriander
Juice of half a lemon
1 tsp garam masala
2 fresh green chillis (optional)

Method

Roast the cumin and coriander seeds in the pan that you will be using to make the curry. Grind in a pestle and mortar and set aside.

Bash the garlic and ginger into a paste with some salt and a drop of water – you can use the pestle and mortar if you like; just make sure you remove the ground seeds first.

Heat some olive oil in the same pan and fry the onions slowly until they are golden. Add the garlic and ginger mixture and fry with the onions for 5 minutes. Add the ground cumin, coriander, turmeric and chilli powder to the pan and fry for a few minutes.

Chuck in the chickpeas, tomatoes and vegetable stock, give it a good stir, and cook for half an hour or so until the sauce has thickened.

Just before serving, add the chopped fresh coriander, the juice of half a lemon and the garam masala and, if you like extra heat, throw in a couple of whole fresh green chillis.

STIR-FRY

Serves 2–4

This is a basic recipe for stir-fry that's quick, easy and cheap. It's a good stand-by recipe and great for using up small amounts of vegetables that are in danger of ending up on the compost heap. There are many variables to this basic recipe and it's a very forgiving one too, so jazz it up according to your personal taste, especially when it comes to heat.

Ingredients

1 tbsp coconut oil
1 tsp Marigold boullion in 100 ml warm water
4 spring onions, chopped (or a small to medium-sized onion)
1–2 chillis (optional)
A handful of cashew nuts
3 garlic cloves, sliced thinly
2.5 cm cube of peeled fresh ginger, grated
A combination of four or five of the following:
Mushrooms (button, chestnut or shitake), sliced or broken into bits
Carrots, thinly sliced
Broccoli, cut into separate florets
Sugar snap or mangetout peas
Baby sweetcorn
Aubergine (I usually like to pre-cook this in a separate pan, searing it in as little oil as possible)
Sweet pepper, sliced thinly
Splash of Bragg Liquid Aminos or soy sauce
Sriracha (Thai chilli) sauce (optional)
Thai basil (optional)

Method

Put all vegetables in a large bowl and pour the Marigold mix over them.

Heat oil in a wok or large, deep, non-stick frying pan. Add spring onions and chillis and stir-fry for a minute. Add cashew nuts and stir-fry for another minute. Add garlic and ginger and stir-fry for an extra minute. Add the whole bowl of vegetables (including whatever liquid is in the bowl) and stir-fry for ten minutes or until the carrots and broccoli are cooked but still firm (al dente). Add soy sauce or Bragg Liquid Aminos (and Sriracha sauce if using) to taste. Throw in a handful of Thai basil, cover and simmer for five minutes.

Serve with brown rice or noodles.

RED THAI CURRY

Serves 2–4

A Thai version of the stir-fry involves a red curry paste which can be made in batches and frozen.

Method

RED CURRY PASTE: Toast cumin and coriander seeds in a small frying pan until smoking and fragrant. Leave to cool.

Put all ingredients, including seeds, into a blender and blitz.

Save in batches of 150–200 g, which can be frozen.

CURRY: Heat oil in a wok or large, deep, non-stick frying pan. Add spring onions and chillis and stir-fry for a minute. Add garlic and ginger and stir-fry for an extra minute. Add 200 g Thai red sauce. Cook for a minute to let the flavours infuse.

Add the coconut milk, vegetable stock, agave syrup and soy sauce. Stir and add the vegetables. Simmer until vegetables are just cooked with a little bite to them.

Ingredients

RED CURRY PASTE:

1 tbsp cumin seeds
2 tbsp coriander seeds
1 medium onion
5 garlic cloves
2.5 cm piece of peeled fresh ginger, roughly chopped or grated
2 sticks of lemongrass
1 red pepper, roasted and peeled
20 g fresh coriander
1 fresh chilli (optional – take out seeds for a milder paste)
3 tbsp tomato puree
1 tsp peppercorns
3 lime leaves
Juice of half a lime
100 ml water (add more if required)

REFER BACK TO STIR-FRY RECIPE ON PAGE 132 FOR SUGGESTED VEGETABLES AND INCLUDE:

2 tbsp coconut oil
2 spring onions, chopped or sliced into threads
2 Thai chillis (optional)
1 garlic clove, sliced
2.5 cm piece of peeled fresh garlic, sliced
400 ml tin of coconut milk
200 ml vegetable stock
1 tbsp agave syrup
2 tbsp Bragg Liquid Aminos or soy sauce

ONION BHAJI

Makes about 12–16 bhajis

This onion bhaji recipe was in my first book, *Our Plot*. They are so good I make no apology for including it again. Like all deep-fried foods they are not particularly good for you, so moderation might be called for.

It takes a bit of practice to get the batter to the right consistency and the oil hot enough so the bhajis are neither over- nor undercooked. If the oil isn't hot enough, the bhajis will be greasy, limp and disappointing.

Ingredients

4 medium-sized onions, thinly sliced
350 g gram flour (chickpea flour)
2 tsp turmeric
Pinch of baking powder
4–6 cloves garlic
2.5 cm nub of fresh ginger, chopped or grated
Some chopped chillies to taste
Salt and pepper
250 ml water
Handful of chopped fresh coriander, including stems
Lemon juice
Sunflower or corn oil for deep frying

Method

Sift gram flour, turmeric and baking powder into a large mixing bowl. Add garlic, ginger, chillies, salt and pepper and mix well. Add water little by little, stirring all the time and making sure all the gram flour is mixed without lumps. The final batter should have a consistency between pancake and cake mix. Err on the side of too thick rather than too thin. Stir in chopped coriander and squeeze in a little lemon juice.

Heat corn or sunflower oil in a wok or a karahi over a hot stove. The oil is ready when a small drop of batter sinks to the bottom, then rises after a few seconds.

At the last minute, add the sliced onions to the batter and begin frying. Work quickly or use small amounts of batter at a time in a separate bowl as the onions will release their liquid into the mix and eventually make it too runny. Place several spoonfuls of the mix into the hot oil and let them sizzle for three minutes or until golden brown. Remove from oil with a slotted spoon and let them drain on a plate with kitchen paper. These are best eaten warm, but can be eaten cold if necessary or reheated in an oven at 160°C for ten minutes (not a microwave as they should retain an element of crispness).

MIXED VEGETABLE CURRY

Serves 2–4

Ingredients

2 tbsp vegetable oil of choice
2 tsp cumin seeds, whole
1 green chilli, whole
½ tsp asafoetida powder
1 onion, sliced
2.5 cm nub of fresh ginger, chopped or grated
3 garlic cloves, sliced
½ tsp turmeric
1½ tsp ground coriander
1 tsp ground cumin
1 tsp chilli powder
2 tomatoes chopped
400 g cauliflower, cut into florets
600 g waxy potatoes, chopped into small cubes
2 medium-sized carrots, chopped
200 g fresh or frozen peas
Handful of chopped fresh coriander
Lemon juice

I never really know what the recipe is for vegetable curry. It seems to come out differently each time but it's based roughly around this template.

Method

Heat 1 tbsp oil in a heavy iron skillet or a deep non-stick pan. Add cumin seeds and chilli and fry for a minute or two. Add the cauliflower, potatoes and carrots and mix well so that the oil and seeds coat the vegetables. Cook for 10 minutes. The aim is to sear the vegetables and not cook them until soft, so if your pot isn't big enough you can do this in separate batches.

Empty the vegetables into a bowl to rest while you make the sauce.

Heat another 1 tbsp oil. Add 1 tsp of cumin and let it sizzle for a few seconds. Add asafoetida; let it fizz for a few more seconds before adding the onion. Lower the heat and cook the onion slowly until it starts turning light brown.

Add garlic, ginger and the rest of the spices to the onions. If too dry add a little water. After a few minutes, once the spices start to emit their fragrance, add the tomatoes. Cook until the tomatoes reduce to make a sauce.

Return the vegetables (except for the peas) to the pan and mix well. Again, add a little water if too dry, cover and cook on a low heat for 10–15 minutes or until the vegetables are cooked through. Add peas and lemon juice and cook for a further five minutes. Garnish with chopped coriander.

Now, here's the thing. Sometimes these curries taste okay immediately after you've cooked them, but more often than not, they benefit from resting for a while. Everyone knows that leftover curry eaten the next day for lunch always seems better than the night before. I always prefer to eat curry when it's cooled down a bit anyway so it's just a matter of timing. Whatever your preferences, experiment a bit and serve with rice, dal, a poori or chapati and your favourite Indian condiments.

MEXICAN MIX

Serves 4

We're big fans of burritos and tacos not just because they are inexpensive, quick to prepare and healthy, but because they are so tasty. You can be as flexible as you like with them, and everything can be made ahead of time. Fillers consist of brown rice, jalepeño peppers, salsa, vegan cheese, lettuce, cucumber, black beans and guacamole. Here are recipes for the black beans and guacomole, which benefit hugely from being homemade.

BLACK BEANS

Method

Fry onions in oil until translucent. Add ground cumin seeds and cook for a minute or two. Add garlic, peppers and yeast flakes. Fry until peppers soften and even brown a little.

Drain beans and add to frying pan and mix well. Crush the beans a little with a potato masher or leave whole and stir in the refried beans (if using).

Stir well, cover and warm through on a low heat for a few minutes before serving.

Ingredients

1 tbsp rapeseed or olive oil
1 red onion, minced
1 tsp ground cumin seeds
1 garlic sliced thinly
1 red or yellow pepper, chopped into strips or small cubes
2 tsp Marigold Engevita Yeast Flakes
1 400 g tin of black beans
½ tin of refried beans (optional)

GUACAMOLE

Method

Halve the avocados, remove the stones, scoop out the flesh and mash with a fork on a flat plate. Leave some lumpy bits to give it texture. Add the rest of the ingredients and mix well. Season to taste and garnish with more coriander before serving. Lime juice will help stop the avocado turning brown; so too will the avocado seeds if left in the guacamole and covered.

Put your fillers in separate bowls, spoon a little of each into warmed tortillas, wrap them into rolls or envelopes, pour yourself a beer, squeeze a little lime and tuck in.

Ingredients

2 ripe avocados
1–2 garlic cloves, minced
2 spring onions, finely chopped
6 cherry tomatoes, chopped
1 jalepeño pepper (otional)
Juice from half a lime
Handful of roughly chopped coriander
Black pepper

RISOTTO

Serves 2

This foundation recipe for risotto can be adapted to whatever vegetables you care to add, such as:

- Pea and mint
- Asparagus and pea
- Mushroom (experiment with a mix of mushrooms)
- Squash (oven-bake chunks of squash with whole garlic before adding to the risotto)
- Beetroot (oven bake beetroots with whole garlic before adding to the risotto – large old beetroots can be a little earthy)

Ingredients

1 tbsp olive oil
1 red or white onion, minced
2 garlic cloves, minced
150 g arborio risotto rice
500 ml warm water mixed with
 2 tsp Marigold bouillion
100 ml white wine
2 tsp Marigold Engevita Yeast
 Flakes
100 g grated vegan cheese
50 ml soya cream
Cooked vegetables of choice
Pepper or nutmeg to season
Parsley, chopped, to serve

Method

Warm stock in a small saucepan.

In a separate pan, fry onions until translucent. Add garlic and fry for a minute. Add rice and mix well with the onions and garlic.

Add a ladle of stock to the rice and stir as it cooks. Adjust heat so that the cooking tempo is neither too slow nor too fierce. Keep stirring! When the stock is used up, stir in the wine and cook for a minute or two.

Add vegan cheese, yeast flakes and soya cream and stir gently to melt the cheese. Add cooked vegetables of choice.

Warm through on a low heat for five minutes, then cover and leave to rest for another five minutes.

Season with pepper or a little nutmeg. Add some chopped parsley and serve.

BAKEWELL TART

Serves 6

This recipe was originally from Emily Cooks Vegan (www.emilycooksvegan.com), but the page has since disappeared.

Ingredients

FOR THE PASTRY:

250 g plain flour
125 g vegan margarine

FOR THE SPONGE:

160 g self-raising flour
160 g ground almonds
160 g caster or granulated sugar
½ tsp baking powder
1½ tsp almond extract
180 ml vegetable oil
150 ml lukewarm water
A handful of flaked almonds

OTHER INGREDIENTS:

Raspberry jam, about 1–2 tbsp. It has to be raspberry for a proper Bakewell tart. However, if you're not bothered about authenticity, strawberry jam is a good substitute.
Icing sugar to decorate

Method

Preheat the oven to 190°C/375°F. Rub the margarine into the flour until it looks like fine breadcrumbs. This will take 5–10 minutes.

Add 3–4 tbsp of cold water until it comes together as a dough. Roll out to about a 3 mm thickness, then line a tart tin with it. Prick the pastry with a fork to stop air bubbles from forming, line with greaseproof paper, weigh down with ceramic beans, then bake for 10 minutes.

Remove your pastry case from the oven and let it cool for a few minutes while you get the ingredients ready for the sponge layer, then spread a generous layer of raspberry jam over the pastry.

Mix together the dry ingredients for the sponge in a bowl. Then add the wet ingredients to it; it will be quite a runny mixture.

Pour it into the pastry case, sprinkle the flaked almonds over the top and bake for 10 minutes, then lower the temperature to 180°C/350°F and bake for a further 25 minutes.

Leave to cool for at least 15 minutes, then dust the top with icing sugar.

Lazy Cat Kitchen also does a fine version of Bakewell Tart with fresh raspberries rather than the traditional jam: www.lazycatkitchen.com/vegan-raspberry-bakewell-tart.

CARROT CAKE

Serves 6–8

Based on a recipe in *The Cake Scoffer* by Ronny Worsey.

Ingredients
225 g carrots, grated
170 g sultanas or raisins
140 g self-raising white flour
140 g self-raising wholemeal flour
170 g golden caster sugar
1 tsp ground cinnamon
1 tsp ground ginger
200 ml vegetable oil
200 ml water
pinch of salt
dash of vinegar
½ tsp vanilla essence

FOR THE ICING
170 g icing sugar
½ tsp vanilla essence
115 g vegan margarine
Cashews or sunflower seeds

Method
Preheat the oven to 190°C/375°F.

To make the cake, stir all the dry ingredients together and then add the wet ones in and mix.

Bake for 45 minutes, then reduce oven to 160°C/325°F and cook for another 30 minutes. Cool in tin.

To make the icing, mash the sugar and vanilla essence into the margarine with a fork. Turn cake out of tin and ice once it is cooled. Optional: top with very roughly chopped cashew nuts or sunflower seeds.

TURMERIC CHAI

Serves 2–3

Ingredients
250 ml water
5 cardamon pods
½ tsp ground cinnamon
2 cm nub of peeled and grated fresh ginger (or ½ tsp ground ginger)
½ tsp turmeric
A turn or two of ground black pepper
Grated nutmeg
300–400 ml plant-based milk (I prefer oat milk)
Agave syrup to taste

Method
Heat water in a pan and bring to a boil.

While you are waiting, crush cardamom pods in a mortar and pestle and grind to a rough powder.

Add ground cardamom, ginger, cinnamon and turmeric to water, grate a little pepper and nutmeg, and boil for three to five minutes.

Add oat milk and agave syrup. Bring back to boil then lower heat and simmer for a couple more minutes before straining through a sieve into cups. Sweeten with more agave syrup if required.

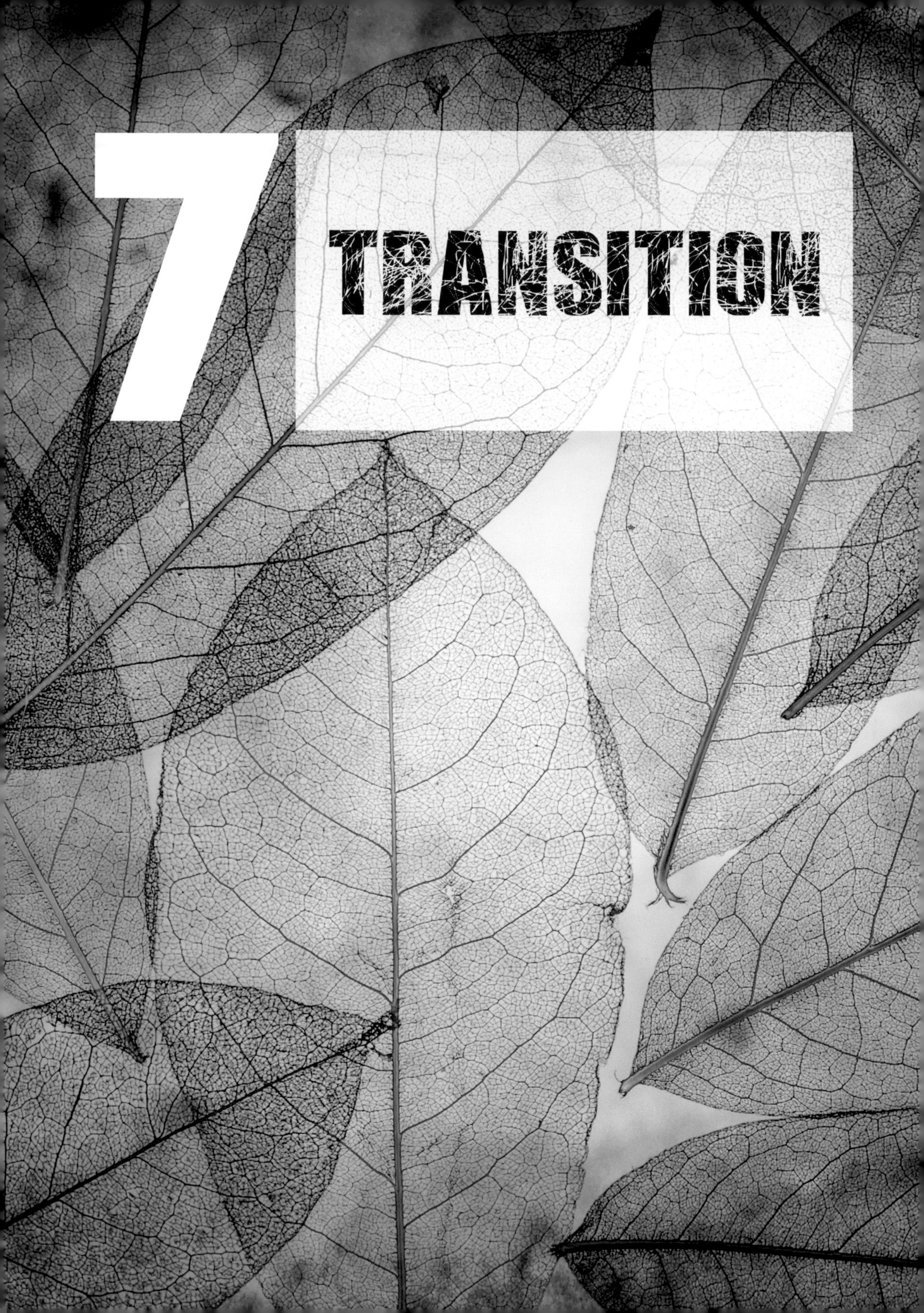
7
TRANSITION

‘WE ARE,
QUITE LITERALLY,
GAMBLING WITH
THE FUTURE OF
OUR PLANET –
FOR THE SAKE OF
HAMBURGERS.’

PETER SINGER, *ANIMAL LIBERATION*

Killed at Newman's Abattoir, Farnborough, on 1 September 2016

TRANSITIONING FARMING

In 2019 a film about a farmer who couldn't face sending his animals to the slaughterhouse won a BAFTA award for best short film. Alex Lockwood's documentary, *73 Cows*, features Jay and Katja Wilde, the first farmers in the UK to give up their entire beef herd and transition to organic plant-based farming. Jay, a vegetarian, had inherited the 172-acre Bradley Nook Farm from his father in 2011, but was conflicted sending animals that he'd got to know to their untimely death. To relieve some of the stress he gave up dairy farming, where the psychological effect on the animals is obvious: 'To take the cow's milk, you have to separate her from her baby. This is really difficult . . . the cows get very upset when they are separated. It takes them a long time to get over it.'

Switching to organic beef so that calves could spend more time with their mothers while maintaining traditional hay meadows was better, but it wasn't enough. No matter how well he looked after his herd, sending them to slaughter was a betrayal of trust. A chance encounter with a member of the Vegan Society provided Jay with a solution to his dilemma by transitioning to arable farming. Stock-free farmer Iain Tolhurst and David Graham from the Vegan Organic Network came to Bradley Nook Farm to discuss the possibilities of producing food without having to exploit any animals at all. When it came to deciding what to do with the herd, Jay just couldn't face sending them to slaughter, and with help from the Vegan Society a new home was found for them at Hillside Animal Sanctuary in Frettenham, Norfolk. Jay and Katja are now vegan and run a stock-free vegetable farm.

It's a heart-warming story for animal lovers and an important example of what can be achieved. It shows us that farmers aren't always as heartless as they might seem to vegans, but are simply doing what their families have done for generations. Breaking free from this cycle, even if our conscience wants us to, can be difficult, but the idea that we can't consider a more compassionate approach doesn't necessarily hold true. All we need is the capacity to think from the animal's point of view. Consider situations where a farm animal has escaped from the lorry taking it to slaughter and spends hours or even days trying to evade capture. Whom do you root for? The animal doing everything they can to stay alive or the people trying to make sure that they end up on the butcher's block? Such stories often make the news, with most people rooting for the animal to find sanctuary. The odd thing is that the majority of these people are also meat eaters and will quite happily go on to eat a steak or hamburger. The reason they can do this is because they haven't made a connection with what they are eating and haven't witnessed animals' desperate fight for survival or the process by which they were killed.

The most famous example is probably Emily the cow, which made the headlines in 1995 by jumping slaughterhouse gates in Hopkinton, Massachusetts, and spending forty days eluding capture. Her escape

Jay and Katja Wilde's farm in Derbyshire

made global news as local residents helped her to evade her would-be captors. She eventually found safe haven at Peace Abbey where she lived happily for eight years, gaining international fame as a symbol for the animal rights movement. A life-sized statue was eventually built on her grave.

While not quite so dramatic, stories like this are not uncommon. Who hasn't seen news clips of sheep being rescued from blizzards or cows and pigs being saved from drowning? They usually serve as feel-good items at the end of a bulletin, yet these same animals will eventually end up on the kill floor. This disconnect between our natural inclination to help the vulnerable and turning a blind eye to violence in order to satiate our desires or cultural traditions is exposed clearly at the Yulin Dog Meat Festival.[1] Advocates saving dogs and cats from being bludgeoned and burnt to death are hailed as heroes, but are branded as criminals and vilified as 'militant' or 'extreme' when rescuing animals from intensive farms where their lives have been a misery from the day they were born. Consider too the public outrage at irresponsible people who leave their dogs in hot cars. Most of us wouldn't hesitate to cause criminal damage to release that dog and save its life. But where's the outrage when we see a lorry stacked full of pigs (generally accepted as being more intelligent than dogs) unable to move, panting for their lives? Why are advocates branded as extreme for giving water to these animals or protesting against live exports when such misery is prolonged over several days?

Another example of such an epiphany involves Devon sheep farmer Sivalingam Vasanthakumar. In January 2019 he found the trauma of

At a vigil outside Tulip Meats, Ashton, Manchester, these pigs were silent. Pigs screaming for their lives in gas chambers can be heard clearly from an adjacent public footpath.

taking his sheep to slaughter too much and drove two hundred miles to a sanctuary near Kidderminster instead. The sheep would have made him good money, £9,000 or so, but the stress of taking them to slaughter and the strain on the sheep themselves finally got to him. Now, after 47 years of animal farming, he grows vegetables and keeps a few cattle to roam free on his land.

Other farmers include Bob Comis in the USA, who gave up 'humane' pig farming to grow vegetables, and Andrea Davis, who turned her goat's cheese farm into a vegan sanctuary. Comis, whose story is told in the 2015 documentary *The Last Pig*, grew to love the pigs he was rearing for slaughter, but eventually his crisis of conscience came to a simple conclusion: 'I don't want the power to decide whether someone lives or dies.' Meanwhile, after nine years in the dairy industry, Davis couldn't stand separating babies from their mothers: 'I came to terms with the fact that there was no right way to do the wrong thing.' Her Sanctuary School at Broken Shovels Farm, Colorado, now takes in abused, neglected or unwanted farm animals and doubles as a vegan education centre. The hope is that as the vegan movement gathers momentum, more farmers will feel conflicted when their actions and morals are misaligned, and will make changes to not only help bring an end to suffering but also steer a course towards a more sustainable way of living.

Breaking away from the engrained norms of animal husbandry and a way of life that has been passed down through several generations will challenge even the most liberally minded livestock farmer. While it's easy for vegans to get upset with farmers who show no compassion for their animals, the truth is that we need farmers more than ever and, equally, they need to see the vegan movement not as a threat but as an opportunity.

Exactly how easy or difficult it is to transition from animal agriculture to arable farming depends on the land and whether the choice of crop can be economically viable and sustainable. It's hoped that studies into how plant-based agriculture can regenerate land and store carbon will result in governments providing support and financial incentives for transitioning farmers, and as more of them make the change (reassured that their livelihood won't necessarily be at stake), others will join them.

Being more self-sufficient while effectively tackling climate change should be at the heart of any agricultural system. Research from Harvard University suggests that the UK could grow enough protein to sustain itself and adhere to the Paris Climate Agreement by turning a proportion of its agricultural land (currently used for grazing and growing animal feed) over to forest and, in doing so, soak up twelve years of carbon emissions.[2] Benefits would include more home-grown food and more forested land for recreation and improved habitat for rewilding projects. The report notes that the UK imports 90 per cent (by value) of its fruit and vegetables and that this, together with the uncertainty surrounding post-Brexit trade deals, could leave the UK in a precarious position.

With half of the UK's land being used for animal agriculture and with low nutritional yields compared to the resource inputs involved, reforesting the UK and growing more efficient crops like beans, pulses and a range of fruit and vegetables provide a good opportunity to supply all our calorie and protein needs while also improving diversity and food security. Regenerated forests will also improve water filtration, flood defences and greater carbon capture, which might just help the UK with its commitment to reducing greenhouse gas emissions (GGE) to zero by 2050.[3]

It's been suggested that if governments are serious about cutting GGE, then meat and carbon taxes are inevitable. It will be interesting to see which political party is brave enough to bite that particular bullet. France's *gilets jaunes* protests against fuel tax in 2018 resulted in social upheaval; the barricading of roads and mass demonstrations eventually led to the plan being scrapped. Fuel taxes hit the public hard, especially those who live outside public transport zones and those in lower income groups when the tax is passed on to them in price rises for food, goods and services. Taxes on meat could also see resistance since it is a perceived 'necessity'. Real change has to come from consumers, and for that to happen we need to understand the consequences of our actions, purchases and lifestyles. As many of our choices are based on convenience, culture and tradition, making changes can seem challenging, but with the stakes so high it boils down to this: do we want to preserve the world as we know it and leave it in better shape for future generations to enjoy? I'm sure that the answer from any reasonable person would be a resounding

'WE HAVE TO DISPLACE OUR SENSE OF ENTITLEMENT IN ALL THAT WE DO, AND HAVE TO START GETTING COMFORTABLE GARDENING WITH A VIEWPOINT THAT IS NOT ENTIRELY HUMAN. . . . WE SHOULD STRIVE TO MAKE LANDSCAPES THAT ARE NOT ONLY ATTRACTIVE AND USEFUL TO US, BUT THAT ARE EQUALLY IF NOT MORE ATTRACTIVE AND USEFUL TO OTHER SPECIES.'

BENJAMIN VOGT,
A New Garden Ethic: Cultivating Defiant Compassion for an Uncertain Future

'yes'. Giving up the things we are addicted to is difficult, but, unlike fuel, meat is not a necessity. Is putting our tastebuds above the future survival of life on earth justifiable? I'm sure that the answer from any reasonable person would be a resounding 'no'.

My own experience as an amateur grower on an allotment has made me appreciate the time and effort that goes into growing food, and Jay Wilde will be the first to tell you that farming isn't easy.

FARMING, LAND USE AND A VEGAN FUTURE

While transitioning farmers provide optimism for a vegan future, many farmers, apart from questioning the logic of growing plants, may find the practicalities involved difficult. Much of our countryside has been shaped by farming practices that have evolved over thousands of years, but much of today's modern approach to agriculture is largely due to the use of fossil fuels and rapid development since the Industrial Revolution. Before considering what a vegan future might look like, it's worth reminding ourselves just how farming evolved in the UK.

It's generally accepted that our agricultural revolution took place between 1750 and 1850 with the increased demand for food from an ever-expanding population.[4] Until then the population had peaked at around 5.7 million because 'agriculture could not respond to the pressure of feeding extra people', but the development of new intensive farming systems (using a rotation of turnips and clover), together with woodland clearing and new fenland reclamation in eastern England, saw a significant increase in crop production. Higher yielding wheat and barley began to replace lower yielding rye, while permanent pasture gave way to more arable land and an increased turnip yield (used for fodder) from land normally left fallow.[5]

By the early nineteenth century farmers had begun to understand the importance of nitrogen. Existing stocks tied up in permanent pasture were ploughed to grow cereals, while manure from animals reared in stalls was used as fertilizer. Nitrogen-fixing legumes such as peas, beans and vetches, which had been grown since the Middle Ages, were also planted,[6] and the culmination of these techniques saw a dramatic increase in food production without compromising soil fertility. In other words, farming in the mid-nineteenth century was, on the face of it, more or less sustainable.

It was short-lived. The advent of chemical fertilizers and energy intensive inputs heavily reliant on fossil fuels soon began to undermine this sustainability. This, together with the enclosure of common land, more efficient machinery and farm management, became the 'agricultural revolution'; the movement of freed up labour to towns to work in factories played an important part in accelerating the Industrial Revolution.[7]

Today in the UK the bucolic vision of rolling hills, hedgerows and meadows is cherished by us all. Like so many, I've been beguiled by the British countryside most of my life. Removing the rose-tinted spectacles, however, reveals a different story. Apart from the embedded suffering that takes place in the animal agriculture industry, its inefficiency and damage to the environment are largely glossed over or ignored. A study in *Global Environmental Change* shows that 85 per cent of the UK's land footprint is currently used to feed animals but provides only 42 per cent of protein and 32 per cent of calories, whereas cereals for human consumption use 6 per cent of the land footprint but produce

POTENTIAL BENEFITS TO THE LANDSCAPE

There is no way of knowing precisely what the landscape in a vegan world would look like, but we can make some educated assumptions:

- **Smaller fields** – an increase in the use of cover crops (green manures) in ley strips. This will play an important role in soil fertility and, where they are allowed to flower between crops, will offset the current lack of flowers where crops are grown. One issue arising from this might include what to do with the large-scale machinery that is currently being used. The transition to smaller fields is not something that will happen overnight, so existing machinery will continue to be used for the forseeable future. The hope is that machines will evolve as new farming techniques develop and that governments will help ease the transition for farmers with subsidies and financial incentives.
- **More hedgerows** – increased habitat and biodiversity, food opportunities from berries, ramial (branch wood) woodchip for fertilizing fields. Ramial woodchip and trees will also be used for fuel.
- **More wetlands** – restoration of marshlands.
- **More market gardens** – smaller farms located closer to cities, fewer food miles.
- **Stock-free systems** – no animal fertilizers.
- **Rewilding** – especially in areas that aren't conducive to arable farming.
- **Less pollution** – cleaner soil, air and waterways.
- **More trees** – woodland, orchards, agroforestry.

approximately 32 per cent of the protein consumed.[8] The study suggests that if meat and dairy consumption were reduced, the land could be used for bio-energy production, forests and biodiversity conservation through rewilding. It won't happen overnight, but a gradual reduction in the demand for meat will force the industry to change from livestock to plant-based products, and over many years farming practices will change. It's already happening and eventually changes to the landscape will be inevitable.[9]

Growing crops to be fed directly to humans (and for biofuels) will not only feed more people but also free some 75 per cent of farmland and offer ample opportunities to offset biodiversity loss, species extinction, soil degradation and carbon loss through reforestation and rewilding. The grazing of ruminants allowed to roam free, as exemplified by the rewilding project at Knepp Farm in Sussex (but without killing them for food), will help to recreate the diverse and dynamic landscapes where biodiversity and a self-sustaining ecosystem can thrive. Even marginal land (not considered suitable for crops) can be turned over to woodland, some leaf crops and rewilding projects.

Landscapes rich in biodiversity such as chalk downland, much of which has been lost to modern farming techniques, could be restored through controlled rewilding projects or dedicated animal sanctuaries managed by organizations such as the National Trust, with extra funding coming from the money saved from the £700 million the government hands out in subsidies each year to industrial farming. Some of this money could also be used to help farmers transition from livestock to arable.

TREES

As a teenager in the late 1970s watching the film *Rollerball*, I was shocked and unsettled by the scene where guests expressed their decadent attitudes in a broken society by setting fire to mature trees at the estate of the host's party. It was way more disturbing to me than the rest of the film where opponents were beaten to death in the sporting arena. Back then, growing up in the heart of Exmoor National Park, my appreciation of trees was more about aesthetics, habitat for birds and a place to escape the dubious attentions of a grumpy ram.[10] Given what we now know about trees, their effect on biodiversity, water quality, drainage, ability to store carbon and make the planet more resilient to climate change, the *Rollerball* scene seems even more outrageous and eerily symbolic of the worst traits of human intervention.

It's not difficult to love trees. We enjoy their diversity and the wildlife they attract. We are fascinated by their form, leaves, flowers, textures, colours and fruit. We climb them, breathe their oxygen, use their timber to make things and keep us sheltered and warm. Some of us even talk to them. We enjoy them heralding the seasons and providing avenues, landmarks, windbreaks and focal points. We celebrate their planting, defend their existence and mourn their passing. They are woven into the very fabric of our psyche, and it's relatively easy to

Tree planting initiatives are crucial when it comes to increasing the carbon storage opportunities that landscapes provide and are vital if we are going to meet the targets set by the UN Paris Climate Agreement in 2015.

understand the importance of trees and their role in the ecosystem, but never have they been more important than they are now.

Tree planting initiatives are crucial when it comes to increasing the carbon storage opportunities that landscapes provide and are vital if we are going to meet the targets set by the UN Paris Climate Agreement in 2015. One such initiative, 'Trillion Trees', is encouraging individuals, charities and organizations around the world to exploit the opportunity to plant 1.2 trillion trees in parks, woodlands and abandoned land. This, together with the trees that could be planted on land reclaimed from animal agriculture, could have a significant effect in offsetting GGE and will be especially important in countries like the UK where only 13 per cent of its land is forested and 85 per cent of its farmland is used for grazing and growing crops for animal feed.[11]

In an attempt to sound mildly optimistic about this, we should remember that only 15 per cent of the UK was forested in the Middle Ages and that it reached a low point of 4.7 per cent by the end of the nineteenth century. In order to increase England's wooded landscape to 12 per cent (currently 10 per cent) by 2060, 5,000 hectares (12,355 acres)[12] of land needs to be reforested each year (that's five to ten million trees).[13] At the moment, we're well short of this figure with only 1,500 hectares (3,706 acres) being planted in 2017 and 1,273 hectares (3,145 acres) in 2018, but a government pledge to put £5.7 million towards planting a 193-kilometre (120-mile) stretch of the M62 with 50 million trees over the next twenty-five years is perhaps a sign that the urgency of the situation is finally being taken seriously.

With the red tape and costs involved in procuring funding and labour for such projects, there are calls from some environmentalists to leave things alone and let nature (animals, birds, wind) do the planting for us. Self-sown vegetation has a better chance of success, with a direct relationship to the peculiarities of the site in terms of weather, fertility and microbial activity in the soil. Cutting out the labour, fuel and materials involved in physically planting trees would also save a huge amount of time not to mention taxpayers' money.

Those who argue against this approach warn of thorny scrub (such as blackthorn, hawthorn and gorse) making an impenetrable redoubt that hinders reforestation as we have come to know it. However, scrubland provides valuable cover for wildlife while protecting vulnerable saplings from grazing mammals.[14] Unlike closed canopy woodland (climax forest) that can inhibit the growth of many of our native trees and shrubs that need light to thrive, scrubland helps to create a more diverse and dynamic landscape, enabling more natural processes to take place. Thorny scrub was once a treasure trove of resources, revered and protected by law, but over many years natural regeneration has been forgotten or misunderstood.[15] Our homogenized version of reforestation with the ubiquitous and unsightly tree guards (occasionally doing more damage than good to the saplings they are meant to protect) seems clumsy, wasteful and far less diverse by comparison.

Quite why the Forestry Commission, the National Trust, the Woodland Trust and other tree planting organizations haven't embraced this method of natural regeneration is puzzling, especially when there are successful models such as Knepp Farm where the natural disturbance of grazing animals helps to perpetuate a dynamic mix of trees, scrub, grasslands and water meadows. Grazing animals negatively affecting vegetative cover in rewilding schemes might be a natural concern for the sustainability of such a system, but botanist and ecologist Arthur George Tansley (who coined the word 'ecosystem') noted that high densities of livestock or wild animals having an effect on the structure of vegetation would be exceptional: 'Red deer (*Cervus elaphus*) and roe deer (*Capreolus capreolus*) are woodland animals which occur in such great numbers in the natural situation that there is an equilibrium between the seedlings that are eaten and the regeneration of the woodland, so that the survival of the forest is not jeopardized.'[16]

Natural regeneration also allows a real sense of place and local distinctiveness whereby plants peculiar to the site or region flourish rather than having a standardized textbook list of trees and shrubs thrust upon it. We must also acknowledge that land ownership plays a big part in how our countryside evolves. One unfortunate consequence, from a vegan point of view at least, is that the biggest incentive for landowners to plant more trees is not necessarily to offset climate change but to provide cover for game birds and the money they can make from shooting them. Woodland creation grants, funded by taxpayers, may be enhancing habitat not for woodland birds but rather for pheasants to be shot. Furthermore, tree planting in these instances may not always be sympathetic to the peculiarities of the topography or respectful of existing views, and after ten years or so, the trees may be grubbed out in order to exploit the funding system again.

While planting more trees sounds like a perfectly good way of storing carbon, planting too many trees or, more specifically, growing trees in the wrong places can have the opposite effect. A huge forest fire that almost wiped out the oil town of Fort McMurray in April 2016, destroying 2,400 homes, was the costliest natural disaster in Canadian history.[17] Investigations showed that the fire was the result of a misjudged campaign to convert peat bogs into timber-producing forest. The peat bogs originally acted as a natural fire suppressant, but plantations of black spruce had drained the swamps, turning wet moss into tinder. The combined effect of burning trees and peat turned a very efficient carbon sink into a huge carbon emitter. Tree spacings to gain maximum growth are now under scrutiny, especially in California where forest fires have become more prevalent.[18]

In China afforestation (growing trees in areas where historically there had been none) using non-natives has caused water tables to

While the template for transition exists, the issue of land rights may hinder any initiatives where agricultural reform is concerned.

drop, increasing the likelihood of future shortages.[19] Even trees that place low demands on the water table can upset the natural balance. In India, seeds of mesquite (*Prosopis juliflora*), dropped from a biplane by a maverick maharaja in the 1930s to green the barren land around Mehrangarth Fort in Rajasthan, have resulted in the tree colonizing large parts of the Thar Desert. Drought tolerant and able to grow deep into the crevices of volcanic rock, this Mexican native has dominated half a million acres of arid land and the Kutch grassland region of Gujarat. While useful for poorer communities in terms of fuel, fodder and flour, it's come at the expense of biodiversity, and steps have been taken by conservationist Pradip Krishen to restore the park around the fort. After various attempts with drills, herbicides and even explosives to get rid of the mesquite's roots (even a small part of which will regrow like bindweed), a team of stonemasons were hired to chisel it out by hand. Remarkably, by locating mesquite roots from the sound the rocks made when struck with a hammer, the craftsmen were able to make effective progress, and the land has now been replanted with native trees that have evolved to adapt to the conditions peculiar to the region.[20]

Reforestation has to take into account the type of tree being planted, how it chimes with the overall health of the forest, the amount of water available and the needs of the local people, especially where the economy depends on the productivity of the land. In areas like California where drought and fire are common, this may even mean the need to remove trees. Overall, though, with ten billion trees being lost globally each year, initiatives that support regenerative ecology through major tree planting projects are vital to sustain a healthy and habitable planet. Protecting the trees we have and restoring them to the places they once grew should be at the heart of every nation's environmental policy. We can also go further and look to agroforestry where trees are incorporated into farming systems.

AGROFORESTRY

Agroforestry is a system of intercropping between rows of trees. The benefits include lower costs, higher yields, crop protection (from the elements, pests and diseases), biomass and increased diversity, which is the key driver for a healthy ecosystem. Biodiversity thrives in woodland edge zones where forests meet open land. Agroforestry, therefore, creates such a zone each side of a strip while at the same time being economically productive with fruit, nuts, flowers and timber.

Stephen Briggs grows wheat, barley and vegetables at Whitehall Farm in Cambridgeshire, 50 hectares (125 acres) of which is managed using agroforestry techniques where apple trees are grown in 3-metre-wide cover-crop strips. He likes the idea of 'tiered' or 'three-dimensional' farming where the space 2–4 metres (6–13 feet) above ground can be exploited to make the land more diverse, productive and profitable.[21] While the size of the field is effectively reduced, the trees and wildflower strips act as windbreaks protecting crops,

As part of an agroforestry system, hazel coppicing can provide fuel, food and increase biodiversity and carbon storage.

reducing soil erosion and improving habitat for insects and birds. It took five years to become fully productive, and the return from the fruit trees occupying only 8 per cent of his land is expected to exceed that of the 92 per cent planted with cereal crops.

Briggs learned much from the late Martin Wolfe, who devoted his retirement years to research and development into agroforestry at nearby Wakelyns Farm. He found that trees didn't compete with crops for water as they sent their roots deeper into the soil, while the trees and cover-crop strips can actually reduce evaporation (compared to bare soil) by 30 per cent. Hazel trees (coppiced and used to heat his farm and research centre) were 30 per cent more productive (being exposed to more sunlight) than a traditional plantation in an orchard or woodland. Biodiversity always wins with an increase in pollinators, and with carbon storage potential there are significant climate change benefits too.

Iain Tolhurst is another grower who is integrating trees into his horticultural system at the Hardwick Estate in Oxfordshire, growing trees for fruit, timber and coppicing (see pages 47 and 56). He's trialling an understory of daffodils, rhubarb and artichokes to increase productivity and profit.

Despite enthusiasm for agroforestry from the Woodland Trust, Soil Association and Royal Forestry Society, farmers and landowners in England have been noticeably slow on the uptake.[22] However, the techniques are gaining traction in the rest of the UK and abroad. With the aforementioned benefits and the overriding concern about the role carbon sequestration will play in offsetting climate catastrophe, it's hard to imagine a future where agroforestry won't be a key player in any comprehensive and sustainable food system.

Once regenerative agricultural techniques (such as reduced tilling and cover cropping) start becoming more commonplace, farmers will increasingly be looking to reduce their dependency on both animal and artificial fertilizers. The stock-free system aims to reduce external inputs to zero. Green-manuring and composting can help sustain such a system; so too will be the use of chipped branch wood (ramial woodchip) and this may affect the structure of the land. Smaller fields, more trees and an increased number of hedgerows will provide a sustainable source of branch wood (and fuel), improving habitat and biodiversity and creating a species-rich patchwork that will negate the need for pesticides. Managing such a system means more local work and ultimately a thriving local community.

SOIL AND ITS ROLE IN CARBON SEQUESTRATION

The idea that trees are beneficial to the environment is a relatively easy concept to grasp. What goes on in the substrate we rely on to grow our trees and the food that sustains us is far less obvious, yet the soil holds the master key to much of the life on earth. Josh Tickell, a celebrated documentary filmmaker and author of *Kiss the Ground* (2017), explains how advances in technology have given scientists the ability to look beyond the macro-layer of life (plants, leaves and roots) to the micro-layer (fungi, bacteria and other micro-organisms) where the possibilities of soil carbon sequestration bring some real hope to the issue of climate change – what Tickell refers to as 'soilutions':

> Plants, through their roots, excrete carbon in the form of root exudates. These exudates feed trillions of micro-organisms, which are involved in extremely complex biological and chemical exchanges with the root systems. In a working ecosystem, carbon excreted from roots is carried through a series of handoffs through the upper, pliable 'labile' layer of soil as it moves into the deeper, more immobile layers of the soil. It is eventually deposited in the form of organo-mineral complexes deep within the recalcitrant fraction of the soil. That's where it can stay for thousands of years.[23]

The idea that we can offset carbon emissions by using more plants will be music to the ears of most gardeners, but we need to go much further than the garden fence. We need to look at the way we grow our food and we need to do it quickly.

Former French agriculture minister Stéphane Le Foll has a plan to put carbon back where it came from, and instead of taking ten thousand years to do this, he wants to do it in as little as two to three decades. He has good reason to be impatient. Oceans help to balance the amount of carbon dioxide in the atmosphere, absorbing it when there's too much and releasing it when there's not enough. The problem is that absorbed carbon dioxide turns into carbonic acid, which in turn increases the ocean's acidity, making life very uncomfortable for phytoplankton. Considering that phytoplankton provide us with 50 per cent of the oxygen we breathe, climate change and rising sea levels (not to mention Brexit and Brentford FC's chances of getting promoted to the Premiership) will be the least of our concerns. Le Foll believes in the restorative power of nature and that embracing regenerative agricultural techniques (large scale, organic, no-till agriculture) is our only real chance of sequestering enough carbon dioxide to slow or even halt the climate catastrophe already in play. It would mean the end of herbicides, pesticides, genetic engineering, monocrops, synthetic nitrogen and intensive animal farming. As Josh Tickell writes in *Kiss the Ground*:

> Call it regenerative agriculture, agroecology, biosequestration or drawdown, the United Nations and most of the world are currently blind to one simple solution to carbon emissions. The irony is that bringing carbon into the soil solves multiple global problems. It reduces carbon dioxide in the atmosphere, it increases fertility of the soil, it helps farmers grow more, it allows the oceans to release the carbon dioxide that threatens to acidify the phytoplankton that produce so much of the oxygen we breathe. (As the concentration of the carbon dioxide in the atmosphere is reduced, the oceans naturally release excess carbon dioxide back into the atmosphere.) In essence, it doesn't matter which of these issues you care or don't care about. It only matters that we align on the single overarching solution in front of us.

Stock-free farmers such as Iain Tolhurst have proved that growing food without the need for animal inputs is economically viable and sustainable. But while the template for transition exists, the issue of land rights may hinder any initiatives where agricultural reform is concerned. Such a topic warrants a whole chapter or even book of its own, so I'll waste no time in recommending an independent report, 'Land for the Many', by George Monbiot and six experts in their field, which seeks to put the inequality and exclusion associated with land ownership back on the political table for discussion.[24] Among other things, the report recommends a revival of market gardening where smaller farms near cities can be leased to gardeners and horticulturists.[25] It also suggests a 'Community Right to Buy' scheme that allows people to take back control of and improve local land: for example, residents in flood-prone towns in the uplands being allowed to buy degraded grouse moors in order to plant trees and restore them as natural flood defences. Current tax laws for landowners would need reviewing and updating, as would the planning system, so that farming and forestry decisions (currently left out) can be scrutinized for the impact they might have on wildlife, ecosystems and food production.

Smaller fields, more trees and an increased number of hedgerows will provide a sustainable source of branch wood (and fuel), improving habitat and biodiversity and creating a species-rich patchwork that will negate the need for pesticides.

AEROPONICS

With stock-free techniques providing a workable template for transitioning farmers, innovative systems that bring food closer to cities to reduce the amount of fuel needed for transport are also needed. Aeroponics, also referred to as vertical farming, is an innovative system where plants are grown in climate-controlled environments up to 30 metres (100 feet) high. Trays of greens stacked high photosynthesizing artificial light and producing up to 70 times more food than traditional field farming might sound like a scene from a science fiction movie, but it's gaining traction around the globe and could well play a part in urban food security.[26] With an incredible 95 per cent reduction in water consumption (from using nutrient-rich misting on roots), plants are grown without herbicides or pesticides in reusable cloth made from recycled plastic. Specialized efficient LED lighting replaces sunlight, and by combining solar/wind power and reusing/recycling heat in a closed environment, vertical farms could eventually provide a net yield where energy is concerned.

One of the first pioneers of vertical farming was microbiologist and ecologist Dickson Despommier, who in 1999 saw the advantages of using less water, less energy and less transport to feed our increasingly overpopulated cities. The concept of growing large quantities of nutritious food all year round without having to worry about labour, the weather or soil fertility seems too good to be true, but vertical farms are already feeding urbanites in the UK, Canada, USA and Singapore.[27] Resistance to vertical farming comes from concerns about the omission of soil in the equation and how the absence of microbes might affect taste. Nutritional value might be a concern as this is dependent on the nutrient solution the plants are grown in. There are also questions about technical malfunctions, human error and the costs of setting up vertical farms in cities where land is at a premium.

Whether or not you like the idea of food being grown in a sanitized disco-lab, the fact that one-fifth of fossil fuel consumption is for agriculture and 80 per cent of the world's population will be living in cities by 2050 should be enough to make us at least consider some more efficient alternatives in food production. Furthermore, with 70 per cent of water being used for agriculture, nitrates from agriculture the most common chemical contaminant of the world's groundwater aquifers,[28] and soil degradation an increasing threat to food security, aeroponics is a revolution in urban food production that will be impossible to ignore. There are limitations on what can be grown in vertical farms (mostly leafy greens), so it's highly unlikely that aeroponics will replace our food systems completely. We can be thankful for that, but such advances in technology and food production may well be a way forward and help to free up land elsewhere that can be returned to its natural state.

LOCALITY

The importance of local distinctiveness has been in the back of my mind ever since an inspiring lecture twenty-five years ago by Sue Clifford about her work with Common Ground, a charity that celebrates the ordinary and everyday things we often take for granted. It champions things that have evolved over many years of culture and tradition and bring a sense of uniqueness and identity to a place, from the most distinguished customs or features to the most subtle and understated. Much of Common Ground's work has been to save orchards and inspire the replanting of new ones where they have been lost to development. Even the tiniest orchards can be enriching spaces that can enhance a sense of place, seasonality and community.

In light of the issues raised in this book and the predicament we could well find ourselves in if our approach to agriculture doesn't change, the allotment where we have gardened now for almost twenty years has taken on a new level of importance. Growing food locally and consuming seasonal food instead of importing it will help to relieve our dependence on imports from overseas. Apart from stock-free outlets, growing our own food is the only way we can be certain that no animals were exploited in the process and that minimum harm was caused to local wildlife.

Being able to buy whatever we want at any time of the year is a real luxury which we ought to wean ourselves from. Apples imported from the southern hemisphere when not available in the UK are a good example. The number of varieties that can be grown in the UK covering a range of seasons means that apples are available from late July to December. Cold storage extends that season to February or March, and then the fruit can be dried, made into chutneys, pureed and frozen for use when the fresh fruit is unavailable.

MARGINAL LAND

In Scotland, Cairngorms Connect has embarked on an ambitious 200-year plan to regenerate 600 square kilometres (230 square miles) of landscape, including the restoration of native woodlands, peatlands, wetlands and rivers. Taking advantage of some of the most spectacular scenery in the country, the vision is to restore a more natural landscape that meets the demands of wildlife and people, slowing water run-off, improving soil quality and carbon storage while providing food, fuel and water for residents and visitors.

In Cornwall, 'Plants for a Future', a resource and database for edible and otherwise useful plants, restored a windswept field of thin, compacted soil to a burgeoning, tree-covered landscape providing shelter, food and habitat for insects, amphibians, birds and mammals. Attracting visitors from around the world, it serves as an education centre that demonstrates self-sufficiency from a wide range of trees and perennials and shows how forest gardening will play a part in farms of the future feeding both humans and local fauna through working with nature rather than against it. The restored landscape is now a nature reserve teeming with wildlife.

Abroad, a great example of land regeneration can be found at Sadhana Forest, in Tamil Nadu in south India, where Yorit and Aviram Rozin transformed 28 hectares (70 acres) of severely eroded land near Auroville back to indigenous tropical dry evergreen forest. The aim was to provide the means for local rural communities to produce their own food and slow the exodus of villagers to city slums. This vision of sustainable living has restored the forests of their forebears while at the same time, through a system of swales, burns and gabions, raised aquifers in the area by an astonishing 6 metres (20 feet). As Aviram Rozin points out, desertification is not just a physical problem; it's a social problem for the people living there.[29] These projects, therefore, not only are good from an ecological point of view, but also introduce faith in the land and faith in the ability of communities to stay together and live a fulfilling life. The success of the transformation at Sadhana Forest, achieved mostly through the work of volunteers, has gained international recognition and fostered sister projects in Haiti and Kenya, and may well serve as a template for other such initiatives where land has been degraded around the world.[30]

While these examples show how poor quality land can be productive, it's worth remembering that a vegan shift would see a 75 per cent decrease in the amount of land needed for growing food, so it's extremely unlikely that we'll have to depend on marginal land for growing food in the first place.[31]

8 ADVOCACY

'THE WORLD IS A DANGEROUS PLACE, NOT BECAUSE OF THOSE WHO DO EVIL, BUT BECAUSE OF THOSE WHO LOOK ON AND DO NOTHING.'
ALBERT EINSTEIN

Killed at Kedassia Poultry, London, on 28 July 2019

If someone had told my teenage self that I would become an animal rights advocate in my mid-fifties, I would have wet myself laughing. Yet once engaged and willing to take on board the numerous issues surrounding animal agriculture, I began to see things from the animal's point of view. The question is this: how do we justify taking the life of an animal with the desire to live when it's not necessary? It's a question we must keep asking when faced with the barrage of excuses and loopholes that arise when questioning the morality of eating animals. This is perhaps the most challenging part of being vegan: letting go of societal norms that have conditioned us to accept animal agriculture and the hidden cruelty that goes with it as normal, natural and necessary.[1]

Seeing graphic footage of how humans treat animals is very hard to watch and, if you haven't seen it before, seriously disturbing. It begs the question: what did you expect it to be like? Like many vegans, viewing the film *Earthlings* and video footage of farming practices in the UK were pivotal in lighting the fire and making me more vocal about animal rights.

I learned about farrowing crates that constrain female pigs for up to six weeks from a week or so before they give birth until the piglets are taken away.[2] I saw the hellish condition that pigs are kept in for the six months before they are killed. I watched the teeth clipping and castration without anaesthetic that piglets endure, the disbudding of calves[3] and dehorning of cows, the lameness in cows caused by incarceration or pelvic damage from calving, and the castrations and tail-docking of lambs, not to mention the roughshod shearing of sheep that can cause wounds that need stitching and occasional fatalities.[4] Lameness that affects more than 90 per cent of sheep in the UK is, it seems, par for the course.

'WHEN THE SUFFERING OF ANOTHER CREATURE CAUSES YOU TO FEEL PAIN, DO NOT SUBMIT TO THE INITIAL DESIRE TO FLEE FROM THE SUFFERING ONE, BUT ON THE CONTRARY, COME CLOSER, AS CLOSE AS YOU CAN TO HIM WHO SUFFERS, AND TRY TO HELP HIM.'

LEO TOLSTOY

The footage that had the biggest effect on me was of the dairy and egg industries. For thirty years I had fooled myself into thinking that no harm comes to cows or egg-laying hens, some of the gentlest creatures on the planet. A relative lives next to a dairy farm, and I now know why the cows bellow for so long, day and night. If the calves are male, they're either killed immediately or raised for a few weeks or months to be slaughtered for veal. If female, they will be reared away from their mother and subjected to the same cycle of forced pregnancies and separations for their short lives. The natural lifespan of cows is about twenty years, but they are killed once exhausted by this process, usually after five or six years. Being strung up by their back legs or forced into a restraining device to have their throats cut will be the very last insult to their short, undignified lives. Like much of the animal agriculture industry, it's a brutal, merciless exploitation of the female reproductive system and motherhood.

Trying to convince people that cow's milk is for cows and not humans is one of the most surreal experiences you can have as a vegan advocate. Some farmers will say that cows like to be milked. The fact is that they are not willing participants in this exploitation. They have been bred specifically to produce more milk than they need and, as a result, need to be milked to relieve the discomfort. They are also fed while being milked so that's another incentive for them to come to the milking parlour more willingly.

Egg-laying chickens are subjected to cramped conditions nothing like the free-range image we imagine when we see the label. There's often more room for a chicken in an oven than a free-range shed.[5] Labels like free-range, organic, Red Tractor and RSPCA Approved are only used to ease our conscience. I used to buy free-range eggs believing that the birds laying these eggs had lived a comfortable and happy life. After just a little research I now know it means nothing. If the chick is male it will be killed after twenty-four hours. They are either gassed,[6] thrown live into a macerator[7] or suffocated in plastic bags. Sometimes they are even set alight.

A chicken's natural lifespan is around eight years and, like other birds, it would naturally only lay 12–20 eggs per year. Today, thanks to poultry breeding genetics, they lay up to 300 eggs per year, which in turn can cause a number of physiological conditions such as egg-

There's often more room for a chicken in an oven than a free-range shed and, for many, the only time they ever get to see the light of day is when they are transported for slaughter.

In *The Sexual Politics of Meat: A Vegetarian Feminist Critical Theory*, Carol Adams suggests: 'Dominance functions best in a culture of disconnections and fragmentations. Feminism recognizes fragmentations . . . In some respects we all acknowledge the sexual politics of meat. When we think that men, especially male athletes, need meat, or when wives report that they could give up meat but they fix it for their husbands, the overt association between meat eating and virile maleness is enacted. It is the covert associations that are more elusive to pinpoint as they are so deeply embedded within our culture.'

binding, egg yolk peritonitis, impacted egg material, osteoporosis, cancer and prolapse.[8] Broiler hens, which have also been bred to grow abnormally large in the shortest amount of time, don't have the required skeleton or the cardiovascular physiology to support this.[9] After just six weeks they can barely stand. Many die before they are stacked high in crates on articulated lorries bound for the slaughterhouse.

THE HOUSE OF SLAUGHTER

Bearing in mind that the word 'humane' means to show compassion or benevolence, how do you humanely slaughter an animal that doesn't want to die? I suppose the easiest way to answer this is to put your companion animal in the place of the victim. Think about it for a bit. Would you take your cat or your dog to a slaughterhouse when the time comes to euthanize them? Of course not. The word 'humane' is another distraction or marketing device that infers something kind or gentle takes place at the abattoir. But, of course, we know different. 'Adding the word "humane" before "slaughter" doesn't cancel out the slaughter. It cancels out your guilt.'[10]

Many people I've spoken to believe that the UK has the highest welfare standards in the world. The film documentary *Land of Hope and Glory* (2017) shows footage from 'high welfare' farms throughout the UK and the ensuing slaughter of these animals. I would strongly recommend viewing this film before making up your mind about whether slaughter can ever be described as humane.

Out of all the slaughterhouse footage I've seen, I don't recall a single instance where the animal was calm before they were killed. Neither have I seen an instance where the animal seemed to be completely unconscious before they met their end.[11] Animals are transported by lorry to a slaughterhouse. Those that go to local slaughterhouses are the lucky ones, as the majority have to be transported many miles and even shipped abroad before being taken to the place where they'll be killed. Rarely will you see a lorry with straw bedding. Most of the time there is just a bare floor, which is often covered in faeces, urine and bile (many suffer from travel sickness en route). The welfare the animals receive up to this point might vary (in most cases animals from smaller farms tend to look better cared for), but they all end up at the same place.

Methods of slaughter vary and are driven more by efficiency, profit and reducing the risk of injury to the slaughterhouse workers than any real concern for the animals' welfare. Getting the animals into the slaughterhouse is not as easy as it sounds as many have already sensed that something's wrong. Any kindness that an animal received from its owner is quickly forgotten as the unfamiliar environment creates confusion and panic. They are prodded with sticks, slapped and shouted at to remove them from the vehicles and into holding pens inside the slaughterhouse, where the sounds and the smell of blood, urine and faeces from their companions will be picked up. If they resist, their tails will be twisted or they'll be poked or kicked. Electric prods are also used. Speed is of the essence as time is money. A captive-bolt gun to the head is meant to render cows unconscious, whereupon they're either strung up by the back legs or locked into a restraining device to have their throats cut. Some cows have to be bolt-gunned more than once, and it's estimated that 5–10 per cent of cows are fully conscious as they bleed to death.[12]

Pigs are often given an electric shock to the head to render them unconscious and keep them still enough to reduce the risk of injury to the slaughterhouse worker. It's estimated that over a million pigs regain consciousness as their throats are slit and they die slowly from blood loss.[13] One of the worst clips of slaughterhouse footage I've ever seen is of pigs who had already been stunned, hung upside down and cut before being dropped into a vat of boiling water to make the removal of their coarse hair easier. Several were thrashing and clearly conscious, dying from a mixture of blood loss, drowning and being boiled alive. It's an image that will haunt me to the grave. While many are sensitive to the use of the word 'holocaust' in reference to what we do to animals, video footage of pigs being killed in gas chambers clearly illustrates why such parallels are drawn.[14] A third of pigs in the UK are killed this way, two or three at a time, forced into small cages that are then lowered into chambers filled with carbon dioxide. As the gas burns their lungs, they thrash and scream for up to thirty seconds before they die.

Sheep are usually electrocuted before being killed. However, in order to comply with halal practices, 1.4 million have their throats cut without being stunned. When inefficient and shoddy stunning is also

LEFT I met both these sentient beings outside a slaughterhouse one day. Both wanted to live. The only difference, in terms of how we treat them, is our perception. **RIGHT** Pigs, like dogs, are clean, social, intelligent animals. A third of all pigs in the UK die in gas chambers screaming in agony.

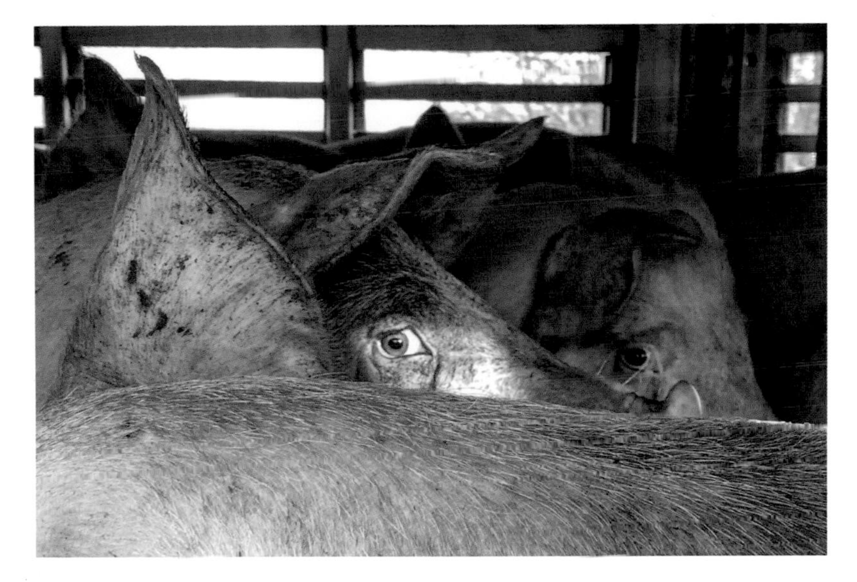

taken into consideration, it's estimated that four million sheep are fully conscious when their throats are slit.

Chickens, turkeys and other birds don't have it any easier. If they haven't already died from genetic disabilities, freezing or heat exhaustion during transport, they are strung upside down on a conveyor to have their throats cut. Halal or kosher slaughterhouses don't allow stunning. Even when birds are stunned (by having their heads dipped into electrocuted water) before their throats are cut, the speed of the production line means that many will be fully conscious during the procedure.

No matter what the animal or the method of slaughter, the complete and utter sadness of this is accentuated by the fact that much of this horror is witnessed by the next in line.

ATONEMENT

The fact that I'd been complicit in such systematic abuse for most of my life stoked a fire I didn't know was smoldering. I'd never considered being an advocate before. Non-confrontational by nature and without the skills for live debate, I knew it was something that I wouldn't enjoy. I wasn't even sure how useful I'd actually be. The problem was how to communicate this information to others, especially friends and family, without sounding like a lunatic.

To begin with I used social media to highlight animal rights issues. Up until then I'd been ambivalent about my presence on Twitter and Facebook, which amounted to nothing more than pictures of plants and gardens, lighthearted observations on football and corny jokes about my dad, but here was an opportunity to show people exactly what was being hidden behind closed doors. I can only imagine what

'THERE IS NOTHING WRONG WITH BEING AN "ANGRY" VEGAN. IF YOU'RE NOT ANGRY, YOU'RE NOT PAYING ATTENTION.'

CONNOR ANDERSON,
@C_Anderson1998, Twitter (3 June 2019)

my followers must have thought when I started posting graphic images of slaughterhouse footage and atrocities to animals. Pictures of pretty flowers were replaced by cows being beaten and sexually violated; gardens were replaced with pigs with their throats slashed thrashing around on the kill floor, and beautiful natural landscapes were swapped for slaughterhouse workers playing football with severed sheep heads. For the first time in my life I was embarrassed and ashamed to be human but, more importantly, I was angry.

Getting people to watch slaughterhouse footage is difficult. Few want to confront the fact that humans are capable of such savagery or accept that they are funding such acts of cruelty. From my point of view, once I'd seen such acts I was complicit by not speaking out. Of course, there are plenty of people who want to keep their heads in the sand and resist change; people are uncomfortable with change at the best of times, and considering the brainwashing we're all subjected to every minute of every day it's hardly surprising.

Facts in the aforementioned films and other independent studies (now freely available on the Internet) are often debunked by those who have an interest in seeing the animal agriculture industry thrive. I too am occasionally sceptical about some statistics quoted by vegan sources, but it's nothing compared to the misinformation we're fed from the meat and dairy industries. Even if only half of the vegan-sourced statistics were true, it would still show that animal agriculture is having a significantly negative effect on the environment and our health and that the way we exploit animals for our pleasure is a gross injustice.

Like many vegans who learn the truth about animal agriculture, processing such information was much more difficult than I anticipated. It's affected some relationships and has changed the way I view the world and our place in it. In some respects, they have been the most unsettling years of my life, yet bizarrely they have also been the most joyous, educated and enlightened. I've seen the very worst of what humans are capable of but also the very best, and based on the notion that one has to remain optimistic, I'm grateful to have this opportunity to raise awareness and speed up the change to a more compassionate world.

Many people think that giving up consuming meat and dairy products is the hardest part of being vegan. It couldn't be further from the truth. The reality is that it's actually quite easy. The most difficult aspect of being vegan is navigating your way in a world that all of a sudden seems disconnected and exploitative and where violence pervades almost everything you interact with. The risk of alienating friends and family is real. It's frustrating and, at times, sad when the people you are closest to won't listen or don't share the same values

In 2018, the UK's first Vegan Garden Festival was held at Hortus Loci Nursery, Hampshire.

of honesty, respect, equality, kindness and compassion: values that you were taught during childhood. It's a price that many vegans pay, but is offset hugely by the knowledge that you are part of a movement that addresses so many of the world's problems and that collectively your small actions can effect positive change, even though it may not occur during your lifetime.

At the time of writing, while many continue to ignore these issues, the horticulture industry is showing signs of promise. In 2018 Hortus Loci Plant Centre in Hook, Hampshire, hosted the first Vegan Garden Festival, and at the FutureScape event at Sandown Park, organized by *ProLandscaper* magazine, I was able to highlight vegan issues during a staged interview with Jamie Butterworth. Jamie inadvertently made it a memorable event by asking: 'Is there a way you'd advise to get into veganism slowly or do you just go cold turkey?' He said it with such innocence it brought the house down. Of course, vegans are all too often the brunt of jokes for their ethical stance. Such jokes serve as a distraction from the serious issues where violence and unnecessary cruelty to animals are normalized.[15] One has to remember that vegans don't take such jokes personally, but are offended by the idea that speciesism itself and the systematic exploitation, torture and murder of animals are made light of. It's tantamount to making fun of racism or sexism.

To be honest, most vegans don't want to keep banging on about everything that is wrong with animal agriculture. The day vegan advocates can get back to all the things they really want to do in life will be a happy one indeed, but most realize that this is a marathon not a sprint, and many older vegans are resigned to the fact that our very last breath before we croak will be 'go vegan!' I've met some of the most dedicated and committed people along the way, and am pleased that the horticultural world is already primed to make

a substantial contribution to animal liberation through its role as stewards of the environment. In time, I hope that more thoughtful and influential gardeners than I am might find it within themselves to make the connection and help speed up the inevitable change.

Some people suggest that there are more important human issues to deal with, but forget that out of the one billion starving people in the world, millions will die while the grain their country produces is exported to western countries to feed cattle and that fifteen million people die each year from diseases caused by eating animal products.[16] Others, especially parents, have told me that they'd go vegan if it wasn't so inconvenient. Vegan parents are quick to point out that it's more inconvenient for animals to have their throats cut than for humans to choose plant-based food. While we all enjoy convenience, do we really want this at the expense of someone else's tortured life and a world which cannot sustain itself?

There are many forms of advocacy. All are effective in their own way, and sometimes it takes a while to find out which ones you are comfortable with and the one that makes the best use of your time. These are some that I've tried to date.

SLAUGHTERHOUSE VIGILS

In December 2010 a group of animal rights advocates founded the Toronto Pig Save by bearing witness to pigs en route to slaughter. This simple act of acknowledgement and the resulting effect on social media platforms where words, photos and videos were shared by advocates have had a profound effect on many non-vegans and vegans looking for a more effective and meaningful way to raise awareness about the plight of innocent animals. At the time of writing, there are over 330 groups in Canada, the USA, UK and Ireland, continental Europe, Hong Kong, South Asia, East Asia, South Africa, Mexico and Central/South America. Though collectively known as the Save Movement, it's unusual for animals to actually be saved.[17] The intention is to bring some dignity to their lives by recording their final moments, acknowledging their existence, and showing non-vegans the reality of what animals are put through when they are treated as products not beings.

I felt compelled to go to my first vigil in September 2016 when a group of advocates from Reading formed Farnborough Animal Action. I remember feeling incredibly nervous during the half-hour journey there. At one point, I almost turned back. I had no idea what to expect, what I would feel and what good it would actually do. Unusually, the slaughterhouse is in a residential road, and occasionally people can hear the screams of animals as they are being killed. Some residents came out to applaud the event, one even going to the trouble of

Activists from Farnborough Animal Action

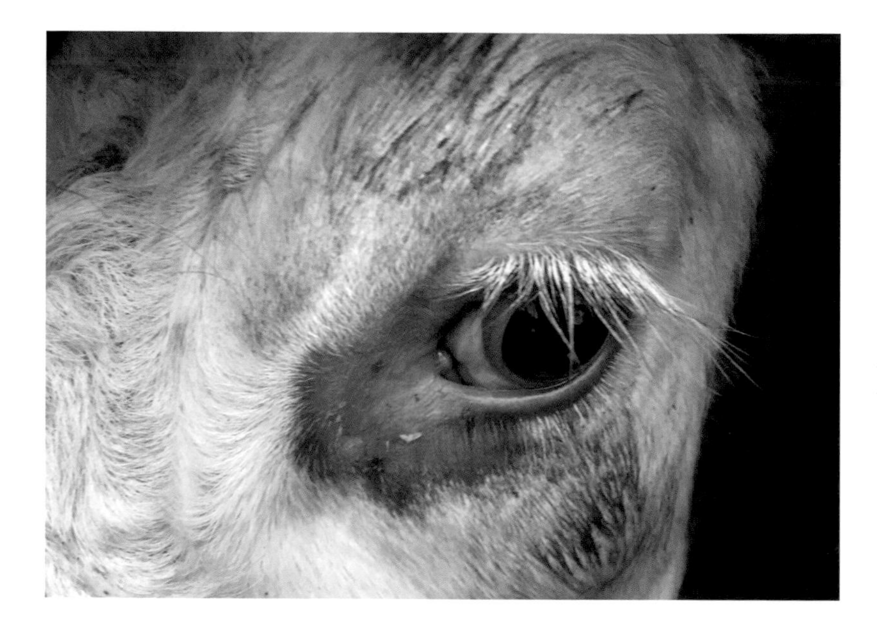

It's only when you look an animal in the eye moments before it meets a violent death that you appreciate the enormity of what is about to happen. Most non-vegans would rather not think of this, as it would put them off their food.

making a pot of soup and refreshments. It was a touching gesture, and despite the classic faux pas made by asking if anyone wanted milk in their tea, it helped to reassure those of us feeling a touch apprehensive about the vigil. It was well attended with advocates coming from as far as Manchester, where the first UK Save event had been held, and it was good to meet other vegans to share stories and experiences. Many, like me, were relatively new to veganism and coming to terms with seeing the world in a different perspective and finding their feet.

When the first lorry of pigs arrived my heart was thumping so hard I could barely hold the camera steady. Looking through the shutters of the trailers, I could see eyes staring right back. Innocent eyes. Confused eyes. Questioning eyes. Eyes that, if you zoomed in close enough, were human eyes. Eyes that you knew. The feeling of injustice was matched only by the desperate sense of being powerless to stop it.

Later I attended a vigil outside a kosher chicken slaughterhouse in east London. By this time I had been going to other vigils for two years or so, and while never easy events to witness, I thought I was mentally prepared for what I was about to see. The articulated lorry delivering the birds had been driven overnight from Yorkshire, and in the early hours we witnessed what can only be described as a living hell. The six-week-old chicks, stuffed into crates piled high, numbered around 6,000. Excrement permeated through the crates to the hapless creatures below. Bred to grow abnormally fast in the shortest amount of time, many could barely stand and some were dead from cold, disease or suffocation. The only time these birds breathed fresh air was while being loaded on and off the lorry on the day of their

Activists attending vigils with East London Chicken Save witness the last moments of chickens, ducks and turkeys at this, the last slaughterhouse in London. Many birds are in such poor condition that they are either dead or dying from the long journey in cramped conditions.

slaughter. To add insult to injury, slaughterhouse workers occasionally laugh at their plight and our attempt to raise awareness. It's sickening to witness such inhumanity, and I'm fairly sure that any non-vegan I know would feel the same.

Why would anyone want to put themselves through such an experience? It's a question we're often asked and there are several answers. First, for me and many new vegans I know, there's a sense of atonement for all the years that we spent ignoring the plight of animals. It helps with feelings of guilt and boosts resolve in continuing the fight for animal rights. Second, and again from a selfish point of view, it's a meeting place for like-minded people joined in the struggle against the oppression of animals. It allows us to catch up with news of other events and discuss ideas for future actions. Third, posting images of victims on social media is a powerful way of raising awareness by showing people the reality of what they are contributing to when they consume the neatly packaged flesh of animals. The meat industry won't show you animals going to have their throats cut because they know it might put people off their food. The reality of putting a face to your food choices literally minutes away from having a knife stuck into their neck is a potent one that does make people think more carefully. Seeing a post from someone you know who has taken the trouble to witness it in person is even more powerful than a random post you might see from an animal rights organization like People for the Ethical Treatment of Animals (PETA) or Mercy for Animals. Overall, these vigils help to expose the truth and strengthen one's commitment to keep fighting for the liberation of animals throughout the world.

BLOG POST: A GARDEN FOR HELL'S GATE, 5 SEPTEMBER 2018

Happy birthday to Farnborough Animal Action, who celebrate their second anniversary this week.

It seems longer. A sign maybe of the intensity in witnessing innocent animals arrive at slaughterhouse gates and looking at them in the eye minutes before their lives are brutally taken.

To mark the event, advocates lay bunches of flowers outside as a mark of respect to the thousands of beings that are butchered there each year.

It's three hours before the first truck arrives. Sheep. The feeling of not being able to help someone about to meet a violent death is impossible to describe, but you see it clearly on the faces of those trying to offer a little comfort to the victims. I nearly drop my phone trying to post a live recording on Instagram and miss my chance before the gates close to the ensuing, respectful silence.

Amidst my fumbling, a message from a friend appears. He says he much prefers my gardening posts on Instagram to my posts about animal cruelty, some of which he can't bring himself to look at. His wife, however, together with his sister and three of his children, have looked at them and are on their way to being vegan. 'Keep up the good work,' he says, 'but please keep gardening.'

After three years of banging on about animal rights, I'm receiving more and more encouraging messages like these. They are priceless. They keep you going as much as the confused and terrified look you see in the eyes of creatures whose only crime is being born with fur, feathers, scales or hooves.

I'm reminded of *The Shawshank Redemption* when Andy Dufresne (writing weekly letters to the state requesting funds for the prison library) finally receives a cheque for $200 and a note saying 'please

A garden at the gates of hell

stop sending us letters.' 'It only took six years,' says Dufresne, 'from now on I'll write two letters a week.'

Yes, there'll be no let-up, I make no apologies for that, but I'll do my best to keep gardening and, in a way, this is a gardening post.

A trailer full of beautiful pigs arrive. A few of them lay prostrate, silent, head down between outstretched hooves, eyes looking up like nervous dogs about to be scolded by their human companions. They know something's wrong. For a brief moment they were someone. Once the gates close they are something; a commodity.

Silence again. Someone instinctively picks up a bunch of flowers and threads individual stems through the Heras fencing. Others follow suit. No conversation, no plan, no worries about plant associations and colour clashes, just heartfelt spontaneity. Any concern that the slaughterhouse workers would take them down is soon forgotten as the bleak industrial fencing becomes a beautiful tapestry. It's clear that no one will be cold-hearted enough to spoil this wall of flowers, at least not while we're here.

This garden, on the very gates of hell itself, is a message of love and hope. It says that veganism is not a fad. It says that those who have made the connection, and can see speciesism for what it is, won't be embarrassed for showing compassion, won't stop banging their drum, posting their pictures or writing their words until the horror we inflict on innocent creatures comes to an end.

It must end for the sake of billions of lives, for the sake of our health, and for the sake of the planet if it is to support life as we know it for future generations.

Change is happening. Be part of it and be on the right side of history. Please leave animals off your plate and live vegan.

The first step to becoming vegan is in acknowledging that animals are sentient beings with an irrefutable will to live their lives without oppression, fear or pain.

INSTAGRAM POST, 23 FEBRUARY 2019

The Don McCullin retrospective at Tate Britain is an astonishing body of work. It's overwhelming in terms of its scope, the atrocities he witnessed, and not only the oppression that humans are capable of inflicting on each other, but also the neglect we are capable of too. Everyone should see it.

It begs the question, how can someone spend much of their life facing such horror and injustice while being powerless to alleviate the suffering in front of them? In the last room the question is answered by McCullin himself:

> I want people to look at my photographs. I don't want them to be rejected because people can't look at them. Often they are atrocity pictures. Of course they are. But I want to create a voice for the people in those pictures. I want the voice to seduce people into actually hanging on a bit longer when they look back at them, so they go away not with an intimidating memory but with a conscious obligation.

These words came flooding back this morning while attending another slaughterhouse vigil where advocates bear witness to animals during their last few undignified minutes on this earth before their lives are taken brutally and unnecessarily.

The pictures advocates post may be a far cry from the technical brilliance of McCullin's and may not have the same sense of jeopardy, but the reason for posting them is the same:

> We want people to look at our photographs. We don't want them to be rejected because people can't look at them. Often they are atrocity pictures. Of course they are. But we want to create a voice for the *animals* in those pictures. We want the voice to seduce people into actually hanging on a bit longer when they look back at them, so they go away not with an intimidating memory but with a conscious obligation.

OUTREACH EVENTS

Outreach events vary in format but usually involve a group of advocates handing out leaflets, providing information about animal rights issues and generally raising awareness. Plant-based milk, cheese and chocolate are often offered as tasters, and occasionally vegan cakes are offered as a reward for watching footage of industrial farming through virtual-reality headsets such as iAnimal goggles.

The Cube of Truth or Anonymous for the Voiceless events are where masked advocates stand in a cube holding laptops or notebooks showing graphic footage of the things animals endure at the hands of humans from birth to the kill floor. The arrangement is immediately arresting to the passer-by, but while the masks attract attention, they prevent any dialogue so the onlooker becomes completely focused

The Cube of Truth or Anonymous for the Voiceless events show graphic slaughterhouse footage in public. Advocates engage with passers-by in an attempt to educate them about things that the meat, dairy and egg industries would rather keep hidden.

on the footage. Warning signs of 'graphic content' allow parents to keep children from seeing such images, but some parents are happy for their children to see the truth. Unmasked outreachers choose their moment to engage with the spectator for a meaningful discussion to take place. Reactions to the images being shown are generally one of shock. Even those who only get a glimpse of the footage while scurrying by look shocked at what they briefly see. The majority of people aren't aware of the processes involved in animal agriculture and the realities of slaughter. While some will repeatedly post 'mmm . . . bacon' to images of animal cruelty on social media, few will actually say this while watching live footage in the high street. As such, it's probably the most effective form of advocacy and probably the most efficient use of one's time, scoring a regular tally of conversions.

DIRECT ACTION

Direct action events vary but are based on peaceful disruption of restaurants, supermarkets and shops where animals products are being sold or consumed. The idea is to bring attention to the violence that has become normalized in our society. While peaceful, they are occasionally met by anger and frustration on behalf of the owners/managers which can sometimes lead to aggression and even violence. Letallbejust says:

> These negative feelings are pre-existing in the people who are confronted by our presence. It is as Martin Luther King Jr said, 'We who engage in non-violent direct action are not the creators

of tension. We merely bring to the surface the hidden tension that is already alive, we bring it out into the open, where it can be seen and dealt with.' . . . We must bring forth the truth of what we do to animals into the forefront of societal discourse so that the normalized violence can be seen, disrupted, ended and replaced with systems that do not exploit and do not act violently without need.[18]

ART

Art doesn't have the spontaneity of a protest, and the effect can't necessarily be measured in the same way as an outreach event where you can accurately document the number of meaningful conversations. The amount of thought and energy that goes into making art, therefore, might seem like a poor use of one's time. However, history has shown us that art plays an important part in any social justice movement. Posters that advertise events or facts, banners, flags and placards at protest marches, films at slaughterhouse vigils, chalking walls or pavements and even 'stories' on Instagram all count towards bringing the animal liberation message to life and creating a different way of seeing things.

The choreographed set-up for the aforementioned Cube of Truth or Anonymous for the Voiceless events is designed along the lines of an installation or a happening, and has proved to be a powerful way of encouraging others to face the things that the meat and dairy industries keep hidden. The most poignant art piece or happening I've witnessed took place at the slaughterhouse vigil in Farnborough where flowers were spontaneously threaded through fencing placed to keep advocates at bay. The fact that the flowers were left on the fencing until the next vigil a month later added even more poignancy to the images that were posted on social media. Did the slaughterhouse owner (who rarely engages with advocates) leave them as an act of defiance or out of respect?

I studied art as part of my degree, had a brief encounter working in the art world when I left college, and Christine is a printmaker, so there's always been a yearning to explore my potential as an artist and, even better, how this might be utilized as a form of advocacy. It's a project for another day, but during discussions with my publisher it occurred to me that Christine could contribute to the book in some way. Each of the chapter portraits is an animal I have witnessed outside the slaughterhouse moments before it is killed. For me they are most personal as they acknowledge the fact that it was someone, not something. Drawing these portraits from photos I'd taken at various vigils wasn't easy for Christine, who has a knack for capturing the character of the animals she draws. My moment with these animals may have lasted just a few seconds, so it was even more difficult for her to really get a sense of whom she was trying to connect with, but somehow she has created a lasting legacy for just a few of the countless innocent creatures killed each year.

Killed at Newman's Abattoir on 22 February 2019

WRITING

Despite the overwhelming nature of social media, the written word is as powerful as it has always been. Poetry, books, fact sheets, musings to accompany Twitter or Instagram posts can make people stop and think more about the consequences of what they consume.

While I was writing this book, Matthew Appleby's *The Super Organic Gardener* was published, the first gardening book dedicated to vegan gardening. It must have touched a nerve. A lot can happen in a year, and by the time this book is published in 2020, I'm sure people will be more open to vegan ethics and consider the subject more thoughtfully and logically.

MEALTIME ADVOCACY

We love to cook and are reasonably good at it, so meals at home are rarely dull. We find when visiting friends who have prepared vegan food for us that it is usually more imaginative and tasty than it was when they cooked for us before we became vegan. Cooking for others, though, chez nous, has become a bit more of an ordeal. Most people, non-vegans and vegans alike, experience a lacklustre meal on a regular basis. It's normal and you forget about it in an instant. However, when we cook a meal for a group of carnists, the stakes are high. It doesn't matter how watertight the arguments are in favour of veganism; if that meal isn't amazing it will be remembered forever and used as ammunition against the idea of veganism.

Meals with non-vegans can be compatible because vegan food can be incredibly diverse and flavoursome. Indeed, some of the best vegan meals I've had lately haven't been at some of the amazing vegan restaurants that have opened over the last few years but have been made by non-vegan friends and family. It can become more challenging at larger gatherings where all preferences need to be catered for, but it doesn't have to be that way. As Melanie Joy says in *Beyond Beliefs* (2017):

> When one person becomes vegan, even though some of the content of eating changes, in many ways the process does not. We can still use our mealtimes as opportunities to honor what they're meant for in the first place: to bond with those we care about. The purpose of rituals and traditions centered around eating is generally not to serve particular foods – fish or potatoes, for example – but to bring people together and create connection. Many families with one or more vegans honor their traditions and mealtimes without causing disconnections, because they realize that the security and connection of the people they love matters more than the one or two ingredients they swap out to create a vegan friendly process. So, for example, they may serve a tofurkey rather than a turkey at Thanksgiving or make their favourite Tex-Mex dish with beans rather than beef.

Of course, there are occasions where the host may not be prepared to make it an all-vegan event. This is something I and many vegans I know struggle with, as it doesn't make for a relaxing occasion at all.

BLOG POST: THE LIBERATION PLEDGE – PEACE BEGINS AT THE DINNER TABLE, SEPTEMBER 2018

There was a time when eating at a table where meat and dairy products were being served didn't faze me at all. Even when I first became vegan I could ignore the smell and the sight of flesh being consumed, let alone the predictable vegan jokes.

Two experiences changed all that.

The first was at an RHS lunch on Press Day at the Hampton Court Flower Show in 2015. Nine plates were served with a very odd combination of pork and beef arranged into a very odd-looking meat tower. Virtually none of it was eaten – too chewy by all accounts. My vegan dish of mange-tout, pea tips, alfalfa sprouts and a balsamic sauce was miniscule by comparison but delicious. As I left I noticed that most of the meat on 20 tables or so (200 covers) was discarded. I came away not only hungry but fuming that animals had to suffer and die for that meal and that their bodies were just tossed away as garbage.

The second was watching the film documentaries *Earthlings*, *Cowspiracy*, *Forks Over Knives* and, later, *Dairy is Scary*. From that moment, being in the company of others consuming body parts or secretions served only to conjure disturbing and, let's face it, perverted images of sexual exploitation and sadistic abuse.

For years I've done my best to avoid such situations and I've been reasonably successful, but of course it comes at a price.

I don't eat with family and friends as often as I'd like and this can also come across as being extreme or disrespectful. Some are understanding, others are confused. A few have occasionally made more of an effort, laying on a vegan spread or ordering meat-free items from a menu, but see no harm in bringing out the cheeseboard or offering milk for coffee. I remember doing the same when my stepdaughter (vegan since the age of thirteen) came to visit. We would always prepare a vegan meal but I didn't see a problem in adding a dash of milk in my tea because no animal had to die for that, right? Wrong, as I've already explained.

When I first became vegan people used to say, 'We'd love you to come to dinner, we'll have some vegan options but we're catering for all personal choices, we hope you understand.'

Once you have seen the truth behind animal agriculture and how they are exploited and killed, it's very difficult to forget or, indeed, explain.

One way of looking at it is this: imagine you are invited to dinner by a friend or family member who has invited a guest from Korea whose preference is to eat dog. Would 'we're catering for all personal choices' justify this and would you feel comfortable sitting opposite someone eating minced whippet or leg of Labrador?

Personally, it would be like sitting beside someone making racist, sexist or homophobic remarks. Again, are you going to feel comfortable? I doubt it. Whether you confront this or not it doesn't make for a pleasant occasion, unless of course you're a racist, sexist or homophobe.

By my reckoning there are four options:

1 Go to the meal, try and ignore the chopped-up body parts and secretions, smile and pretend that everything is okay.
2 Go to the meal and use the opportunity to talk to everyone there about the reality of animal agriculture, the harm it does to animals and the environment, and the negative health implications.
3 Don't go to the meal.
4 The host makes a vegan meal for everyone.

From experience, people naturally prefer you to choose option one, get annoyed at option two, and are either relieved or disappointed at option three. Option four is by far the easiest but is rarely offered. Catering for those who prefer violence in their food is, it seems, more important.

I was pleased, therefore, to hear of the Liberation Pledge. It's a mild form of advocacy but is also useful for self-preservation. Feeling physically sick or judging people watching them consume their meal in close proximity doesn't make for a good experience, often bringing flashbacks of graphic footage of animal abuse. Some can ignore it, others can't. I'm in the latter camp and will go to great lengths to avoid such situations.

The Liberation Pledge encourages vegans to do the following: publicly refuse to eat animals. Publicly refuse to sit where animals are being eaten. Encourage others to take the pledge and wear the symbol for the pledge: a fork bracelet.

Unknowingly, I had already taken the pledge but had never really considered it as a form of advocacy and, having never worn an item of jewellery in my life, was curious about the bracelet.

While the bracelet is a useful conversation piece, the Liberation Pledge is also a form of self-preservation. Advocates see much of the horror that is hidden from consumers for obvious reasons, so confronting it again at mealtimes, which are meant to be enjoyable, relaxed and peaceful, instead becomes at the very least uncomfortable, often stressful and occasionally traumatic. Even with close friends and family whom you know to be good people, it's hard not to be judgemental about their insensitivity towards the exploitation of and cruelty to animals for a fleeting taste sensation. This doesn't make for a happy occasion, and most reasonable people wouldn't want their guests to go home disturbed by the experience.

While the Liberation Pledge focuses on the dinner table, I also tend to avoid buffets or events where canapes of pig flesh are being shoved under your nose every minute and people get uncomfortably close trying to make themselves heard in a crowded room. Dodging

The Liberation Pledge bracelet

fish breath or projectiles landing in your drink really doesn't make for a pleasant occasion.

Apart from unexpected, work-related situations, the only exception I tend to make these days is at a funeral reception. People are more preoccupied with the death of a loved one than the death on their plates, and a rant about animal rights won't endear you to non-vegan mourners. Ironically, having watched 'What the Health' and many nutritional videos by Dr Michael Greger, it's clear that the death on their plates might well have caused the death of the loved one in the first place. Actually, it's probably best not to mention that either if you want to get out of there alive.

My first attempt at making a Liberation Pledge bracelet failed miserably. Finding a fork that I liked was much harder than I anticipated. Bending the one I eventually settled on was even more of a problem and I had to ask my good friend Bamber Wallis to make it for me. Bamber is an artist-cum-water technician, and as he had helped me with water features on several projects and show gardens, I knew he'd have all the tools, but as it turned out even he had trouble and had to resort to heating the fork with a torch in order to bend it. What I thought was going to be a ten-minute job turned out to be almost an hour of hard graft. If he'd known how tough it was going to be, I'm sure he would have told me to fork off or where I could insert it . . . sideways.

Jewellery isn't really my bag, and what with it bashing around on the drawing board, snagging on pockets and hooking sensitive parts of the anatomy while bathing, my patience with it came to an end after a few months. I still have it but the plan is to swap it for a tattoo (something else I've never really considered having before), which is less cumbersome and annoying. Most appealing is that a tattoo might also be useful protection from anyone trying to force-feed me flesh or secretions if ever I have the misfortune to be hospitalized or unable to communicate in my dotage!

Other forms of advocacy include single-issue demos such as hunt sabotage and protests against horse racing, animal testing, zoos, donkey rides and circuses, to name just a few. Protests against live exports at ports where animals are shipped abroad to be slaughtered (often after many hours travelling in congested lorries) are also becoming more newsworthy, as is the plight of Dartmoor ponies being shipped to Europe to be slaughtered for their leather.

Being vegan doesn't just involve avoiding meat and dairy products. Animals are experimented on, used for entertainment, exploited for their fur and used as beasts of burden, so any product or organization that exploits or mistreats animals in any way is boycotted.

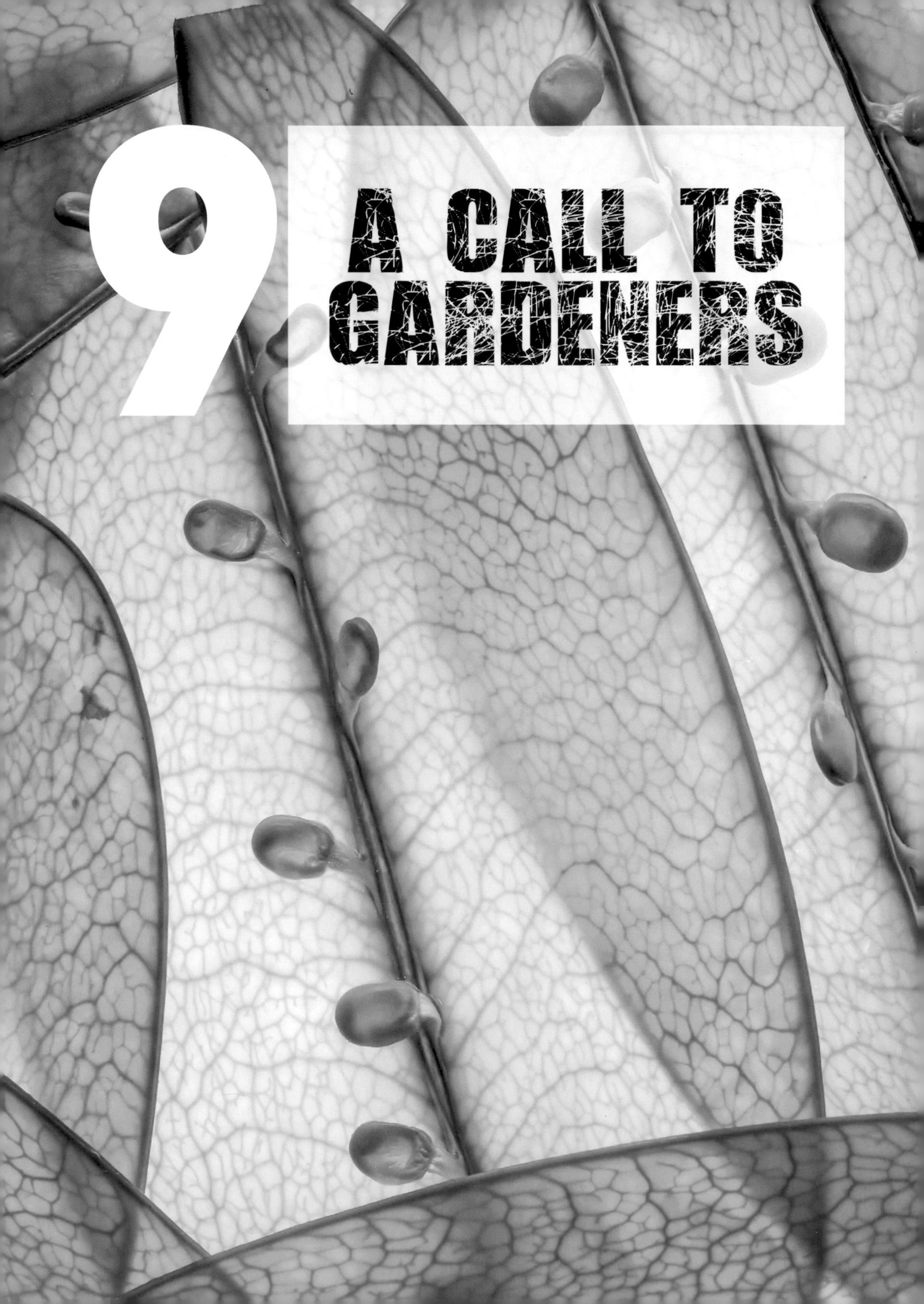

9 A CALL TO GARDENERS

‘PLANTS ARE NATURE’S ALCHEMISTS, EXPERT AT TRANSFORMING WATER, SOIL AND SUNLIGHT INTO AN ARRAY OF PRECIOUS SUBSTANCES, MANY OF THEM BEYOND THE ABILITY OF HUMAN BEINGS TO CONCEIVE, MUCH LESS MANUFACTURE.’

MICHAEL POLLAN, *THE BOTANY OF DESIRE: A PLANT’S-EYE VIEW OF THE WORLD*

Killed at Newman's Abattoir on 23 May 2017

OPTIMIST OR PESSIMIST?

The seed for this book was sown, rather bizarrely, at the wake of the late, great landscape designer John Brookes MBE. Widely regarded as 'the man who made the modern garden', many students, including myself, had the good fortune to benefit from his wisdom, mentorship and acerbic wit. He was a dear friend too, and his sudden departure just a few weeks after his memoir was published was a great loss to the world of landscape design and to those who loved him.[1] John's publishers, Jo Christian and Gail Lynch, introduced themselves after the service. I'd written the foreword for John's last book and they wondered if I had any plans for a book of my own. Recalling how infernally long it took to write *Our Plot*, my immediate response was a flat 'no'. But when they asked me to think about it and whether there was anything in particular – anything personal – I'd like to write about, I wondered how much a book on veganism would make the shelves of the gardening section creak and groan.

Garden writing is hallowed ground where good taste, culture and a modicum of quirkiness are embraced. It's a comfort zone I enjoy myself and not one to be sullied by the darker side of humanity, the destruction it causes and one's own role. I was reasonably confident, therefore, that they'd reject any proposal I had in mind, so on pitching *The Garden of Vegan* over lunch, I didn't hold back. Jo and Gail listened patiently to my story about what prompted me to adopt a vegan lifestyle: the shock of seeing the slaughterhouse footage, the information I had absorbed about animal agriculture and its effect on health and the environment, and the realization that I was working in an industry that might just have some of the answers to some serious global problems. To my amazement (and more than a little panic), they said, 'Fine, when can you start?' In an attempt to sugarcoat the underlying message, I promised to offset some of the inevitable doom and gloom with a little humour, my first thought being that John would have been very proud of the fact that I'd more or less sealed a book deal at his funeral.

The reality though, as anyone who has made it this far will have realized by now, is that I've come up a bit short on the humour side, so to offset this shortcoming, here's a joke:

Q. How can you tell there's a vegan in the room?
A. Don't worry, they'll tell you.

You're bound to have heard it before, it's an old one, but probably older than you might think. Replace the word 'vegan' with 'suffragette' and you get the idea.

The point is that people who are against injustice are rarely quiet. Yet I've been told several times by various people that I'd convert many more people to veganism if I didn't bang on about it so much.

Of course, no one has managed to explain to me exactly how that works. The only reason to make such a suggestion is to stop vegans from questioning the morality of our food choices that, in turn, fund animal cruelty. What's odd is how people who know I'm vegan like to bang on about their hobbies or eating habits that involve cruelty to animals as if to say, 'I know you spend a lot of time campaigning for animal rights, but nothing's going to stop me doing the things I like to do because my enjoyment is more important than their lives.'

With a few chapters under my belt it was clear that there was little to laugh about. Aside from the exploitation of and cruelty to animals, the environmental ramifications were off the charts. I already knew that things were bad, but the more I researched the bleaker the outlook. In what I think was probably desperation, a friend (who has to put up with my 'the end is nigh' rhetoric on a regular basis) kindly bought me the late Hans Rosling's optimistic bestseller *Factfulness*, no doubt to protect my sanity (and hers!) more than anything else. Rosling attempts to reassure the one billion who live in the relatively comfortable income bracket earning over $32 a day (level four) that things really aren't as bad as they seem.[2] He suggests that human beings are evolutionarily primed for bad news, and our perception of the world is all too easily skewed towards pessimism through the misinterpretation of statistics and commonly held beliefs. Statistics he uses show that, in terms of living standards, wealth and the number of people killed in conflicts, things are improving and we are living in the best of times. He encourages us not to worry about everything but embrace a worldview based on facts, as this will help us focus on things that threaten us most.

However, Rosling deftly skirts round acknowledging some of the less tangible (but more serious) threats, mentioning climate change only in passing and completely ignoring the threat to biodiversity and the potentially catastrophic decline of insects. Cherry-picking ambiguous facts suggesting that black rhinos, tigers and giant pandas aren't quite as endangered as we might think ignores not only the World Wildlife Fund's (WWF) more sobering assessments, but also the fact that human activity is responsible for the current sixth mass extinction and that the loss of the smaller, less obvious species can have disastrous consequences for the ecosystem.[3]

Rosling's blinkered optimism is concerning. His anthropocentric bias and assurances that 'things aren't as bad as you think' pay little heed to important statistics on climate breakdown, species extinction, ocean depletion, deforestation, soil degradation, desertification and the rise in plastic waste. This head-in-the-sand approach serves as a useful diversion for anyone who'd rather not think about what the future might hold for life on earth and what they can do to offset it. The fact that it's been lauded by the rich and privileged whose lifestyles threaten those living in relative poverty (levels one to three) perpetuates the myth that all is right with the world. Little wonder it's a bestseller.

'AS FAR AS EATING IS CONCERNED, HUMANS ARE THE MOST STUPID ANIMALS ON THE PLANET. WE KILL BILLIONS OF WILD ANIMALS TO PROTECT THE ANIMALS WE EAT. WE THEN DESTROY OUR ENVIRONMENT TO FEED THE ANIMALS WE EAT. WE SPEND MORE TIME, MONEY AND RESOURCES FATTENING THE ANIMALS WE EAT THAN WE DO FEEDING HUMANS WHO ARE ACTUALLY STARVING. THE GREATEST IRONY IS THAT AFTER ALL THE EXPENSE OF RAISING THESE ANIMALS, WE EAT THEM AND THEY KILL US . . . AND INSTEAD OF RECOGNISING THIS INSANITY, WE TORTURE AND KILL MILLIONS OF OTHER ANIMALS TRYING TO FIND A CURE, A CURE TO THE DISEASES CAUSED BY EATING THE ANIMALS IN THE FIRST PLACE.'

MIKE ANDERSON, author of *The RAVE Diet and Lifestyle*

Thankfully, the esteemed biologist (the Darwin of our times) and National Geographic Society's Hubbard Medal winner E. O. Wilson stopped my cynicism from getting out of control. Wilson believes that despite our greed, shortsightedness and the tendency towards warring tribalism which renders us useless when it comes to making long-term decisions, there is still hope and suggests that half the planet should be set aside for conservation to help conserve 80–90 per cent of all species.[4] Bearing in mind that animal agriculture currently occupies nearly half the global surface area (mostly for incredibly small yields that a vegan shift could put to shame), this can only happen 'if people eat significantly less meat and livestock products, a trend that is currently going in the opposite direction globally'.[5]

Wilson thinks it scandalous that since Carl Linnaeus (father of taxonomy) started cataloguing plants in the eighteenth century, only two million species of the estimated ten million or so life forms have been recorded. But despite this frustration he remains optimistic that his plan is achievable and that the techno-digital age will help us become more efficient, or 'get the most from the least':

> I think one of the main reasons we haven't moved more quickly is that we haven't made that moral shift where people value what they call wildlife. Not just the big animals, but the rest of life. Once people get a sense of what that is and how important it is for nature's ecosystems and, ultimately, for them, I believe

> a shift towards the preservation of the rest of life would follow, to our all-around benefit.[6]

This shift can't come soon enough for the billions of sentient beings we perceive as food and the predicted environmental catastrophes caused by animal agriculture. Whichever way you look at it, industrialized farming is sucking this fragile planet dry. Whether you eat animals or not, we know this to be true. The question is why we spend so much of our time trying to debate the subject, looking for loopholes to justify something we know is causing harm, instead of actually doing something about it. Anyone who has a genuine concern about the environment will know by now that animal agriculture contributes enormously to climate change and that the most effective way to address the situation is to shift to a plant-based diet. The beautiful thing here is that no governments have to be lobbied, no laws have to be changed, you just stop buying it. It couldn't be easier. So why is it so difficult?

VEGANISM: A PARADIGM SHIFT

The answer lies in the belief system in which the majority of us share, something that Dr Melanie Joy, professor of psychology and sociology at the University of Massachusetts, Boston, refers to as 'carnism',[7] a paradigm that enables us to feel comfortable eating cows, chickens, pigs, and sheep but disgusted at the thought of eating other animals despite their anatomical similarities and recognizable sentience.

Those who consume animal products (including vegetarians) use what she calls the "three Ns of Justification – normal, natural and necessary'[8] – to justify their behaviour and often use this to invalidate the compassionate system of veganism. Her work also examines the psychology of oppression and how dominant belief systems can feed injustices where race, gender and sexuality are concerned, concentrating power in one group at the expense of others. Her belief is that the oppression of living creatures soon leads to feelings of superiority that will inevitably be forced on to other humans.

Breaking away from this belief system can be challenging and life changing. Psychologist Clare Mann argues that veganism challenges the social norms the majority of us live by as if in a trance-like state:

> Becoming vegan requires an individual to change their thinking and actions in ways that affect them every day. Each day the vegan makes choices to seek out products and services that don't directly or indirectly abuse animals, and this takes time and energy. For the ethical vegan this is not a problem but for someone not distressed by animal use it can be seen as an inconvenience and of little importance.[9]

A paradigm shift from carnism to veganism often has a profound effect on the way one views societal norms. Those who adopt a vegan diet for environmental or health reasons may not feel so outraged at the violence and suffering involved in animal agriculture. Some are even reluctant to use the term 'vegan', preferring 'plant-based' instead. Outlets recognizing the opportunities from the vegan pound are still nervous about upsetting their regulars, as we found while on a short break in the Kent countryside. A friend had recommended a café which had some great reviews and tasty vegan options. On arriving, I asked the waitress what vegan food was being served and got a smiley but ever-so-slightly curt response: 'We call it plant-based here.' Twice I said the 'V' word and twice I was 'corrected'. The food was good, so ignoring the ticking-off we returned the next day. 'Oh, hello again,' she said brightly and with no prompt from me, 'let me tell you about today's vegan options.' It turns out that the 'V' word is not so difficult to say after all.

A friend enquiring about progress with the book asked if I was going to stick with 'vegan' or refer to it as 'plant-based'. Aside from the fact that 'The Garden of Plant-Based' doesn't quite have the same ring to it, there's a dishonesty in referring to veganism as 'plant-based'. Of course, I do understand that 'plant-based' is less confrontational than 'vegan', but there's quite a difference so forgive me for reiterating the definition of veganism: 'A philosophy or way of life that seeks to exclude, as far as practically possible, all forms of cruelty to and exploitation of animals for food, clothing or any other purpose.'[10] Plant-based is an option for flexitarians cutting down on meat (like Meat Free Mondays). It's a step in the right direction. It would reduce the numbers of animals being farmed and might still have a positive (if much reduced) effect on mitigating climate change, but it doesn't have the interest of the animals at heart.

The thought of accidentally consuming something with animal products in it is enough to upset any vegan, whereas someone who's plant-based may not see it as a problem at all. Many vegans can't stand the idea of others bringing animal products into their home or workplace; neither would they buy a round of coffee with cow's milk because that would clearly be against their principles of funding cruelty in the dairy industry. Someone who's plant-based, however, won't consider it an issue. Once you've made this connection with being vegan, there's no going back.

PESTS AND VERMIN

For all the advice we get about ridding the garden of aphids, weevils, slugs, moles, rats, squirrels and pigeons, it turns out that the most serious pest of all is us: *Homo sapiens*. Latin for 'wise man', it turns out that the inferred wisdom is flawed. Unless there is a major shift in moral consciousness, selfishness may well be the ultimate downfall of our species; the only real calamity would be the number of other species we take down with us.

It brings to mind a quote from the 1999 film *The Matrix*. Near the end of the film Agent Smith says to Morpheus:

> I'd like to share a revelation that I've had during my time here. It came to me when I tried to classify your species. I've realized that you are not actually mammals. Every mammal on this planet instinctively develops a natural equilibrium with the surrounding environment. But you humans do not. You move to an area and you multiply until every natural resource is consumed and the only way you can survive is to spread to another area. There is another organism on this planet that follows the same pattern. Do you know what it is? A virus. Human beings are a disease, a cancer on this planet. You are a plague. And we are the cure.

There's no denying that our presence is detrimental to the planet.[11] We spread more disease and cause more destruction than other animals. Unlike the unfair label we cast upon creatures just doing their best to survive, we are the very epitome of the word 'vermin', and our success as a species has come about at the expense of much suffering.

We use resources more for enjoyment than survival. Balance and moderation is the only way to give future generations a fighting chance to enjoy the sort of lives we currently lead. The only hope is that the intelligence and 'good' within us can see beyond profit and personal gratification. On a more positive note, only relatively recently have we become aware of our negative impact on the planet, and there are many organizations trying to galvanize support for future generations.

Living in moderation has to be the answer, but with all the trappings of capitalism and with our inherently selfish and selective empathy, our ability and willingness to turn things round for the better will be challenging. Governments must act but if history has taught us anything, it is that we, as consumers, can effect change more quickly. If we don't we'll become the generation that could have done something but didn't, and history will judge us as the ones who stood idly by for hedonism and convenience.

A CALL TO GARDENERS

The motivation behind writing this book is that it might inspire gardeners and horticultural organizations with more reach and influence to get behind this important movement. Perhaps I'm being a little naive, but as most of the gardeners I've met over the years have been kind, decent people who are passionate about plants and the wider landscape, is it not reasonable to think that they would be at least sympathetic to the idea of protecting the environment? Is it unreasonable to think, based on the evidence that animal agriculture is so destructive on so many levels, that these same people would, at the very least, consider changing to a plant-based diet?

Not all gardeners will have the time or inclination to look beyond the garden fence, but those who do will feel unsettled by what they see. Some of the problems that the world is facing can seem overwhelming and insurmountable, but if we all unite behind the one thing we all have in common – plants – we can make significant changes and improvements. Much depends on our sensitivity towards nature, our understanding of how every living thing is connected, and how a respect for all life forms is a big step towards reparation. It should arm us with all we need to help reset our moral baseline and steward this gem of a planet in a responsible and compassionate way.

As more people shift to a vegan lifestyle, the less likely it will be seen as extreme. Keeping a low profile at the Chelsea Flower Show 2018, I was pleasantly surprised by the number of people who came to talk to me about veganism. I arrived wondering whether those who'd stumbled upon my social media accounts would want to avoid me, but many initiated the conversation and were genuinely interested in the issues. While not committing to going 100 per cent vegan on the spot, some gave every indication that they would make an effort to do some research of their own.

The Royal Horticultural Society has made tentative steps towards embracing veganism, with vegan food outlets at its events and allowing a vegan-inspired conceptual exhibit at one of its shows. But its decision to accept an application from a well-known dairy farm to sponsor a show garden at the 2020 Chelsea Flower Show was disappointing. With organizations such as the Society of Garden Designers and the Landscape Institute looking to veganize their events (acknowledging the negative impact of animal agriculture on climate change), my hope is that the RHS will one day do the same, and encourage its members to be at the vanguard of the movement, with shared goals of compassion, fruitfulness (food security) and sustainability.

Taking responsibility is key to success. As a child I knew of environmental problems that we'd been made aware of at school, much of it about the pollution of our rivers, which we could see for ourselves. When it came to writing an essay about potential solutions, I didn't have any useful answers and suggested that the agriculture and manufacturing industries should have the technology and bear the cost of cleaning up the rivers they were polluting. After all, what could I do as a schoolboy?

The ecological crisis we face today has much to do with thoughtless consumerism. We are the culprit and we know that we have a responsibility to do something as individuals to leave our environment in a better state than we found it. Time, though, is of the essence. Advances in medicine and science mean that we have seen a significant rise in the human population in the last hundred years. With the world's population set to grow to 9.1 billion by 2050, the demand for resources will rise exponentially. The need for food will rise by 70 per cent, increasing the demand for energy and putting pressure on water supplies, especially for the 1.4 billion people living in areas with sinking ground water levels.[12]

The problem is that we are using the world's natural resources 1.7 times faster than the planet's ecosystems can replenish them.[13] So, the question is, how can we speed up the reparation? How can we inspire gardeners to understand that we, with our fascination, passion and curiosity for everything this world can offer us, can make a difference? It's all very well for this garden designer to lay bare his frustrations in a book, but he knows he has limited clout in the world of horticulture. He knows that the vegan agenda is so taboo that many of his friends and family, even the ones who buy this book, won't even read it. His hopes lie with the more influential horticulturists who can help spread the message far enough for larger horticultural bodies to take the lead and encourage their members to make the necessary changes to their eating habits. You may think Brentford FC has a better chance of winning the European Cup. Maybe. But when you consider the seriousness of the situation and what's at stake, anyone with a genuine appreciation for the world we live in will realize it's worth a try.

RHS AND RGB KEW

As I've already mentioned, one of the largest organizations that could make a significant difference is the Royal Horticultural Society (RHS). The other is Royal Botanic Gardens Kew (RGB Kew). It seems obvious to me that they should embrace veganism simply because of their shared USP: it's all about plants. While the fear of alienating the majority of their members and supporters is clearly a concern for them, the imminent threat of climate breakdown and the ecological repercussions that all life on earth will suffer as a result of inaction ought to be more than enough in terms of motivation. With a United Nations report (8 August 2019)[14] suggesting that a global shift towards a vegan diet is one of the most significant ways to reduce greenhouse gases from the agriculture sector and the EAT–Lancet Commission on Food, Planet, Health (16 January 2019) recommending a plant-based diet to safeguard the future,[15] it's odd that both RGB Kew and the RHS remain so silent when it comes to recommending a plant-based diet.

To be fair, I should point out that both are aware of climate change and the associated environmental implications. A statement on the RHS website acknowledges: 'Climate change is likely to be one of the defining challenges of the 21st century and how we respond will not only determine our future prosperity, health and wellbeing, but also the sustainability of earth's natural environment.' The RHS should also be commended for its decision to allow Joseph Gibson's show garden 'Conscious Consumerism' (highlighting the fact that the meat industry is responsible for up to 91 per cent of Amazon destruction and asking questions about the consequences of our consumer-driven lives) to be built at the Hampton Court Flower Show in 2018. The tendency, though, is to concentrate mostly on informing members how to deal with (or even exploit) the expected changes rather than how to reduce consumption of the things that are causing it in the first place. Encouraging people to cut back on their consumption of fossil fuels and meat to offset climate change might confuse their overriding mission statement: 'The encouragement and improvement of the

Joseph Gibson's show garden, 'Conscious Consumerism', at the RHS Hampton Court Flower Show in 2018

science, art and practice of horticulture in all its branches.' But if the situation is as critical as UN scientists predict, then paying lip service to a plant-based lifestyle by allowing a show garden and a couple of vegan food outlets at a horticultural event clearly isn't enough.

RGB Kew's mission states: 'We want a world where plants and fungi are understood, valued and conserved – because our lives depend on them. RGB Kew's mission is to be a global resource for plant and fungal knowledge, building an understanding of the world's plants and fungi upon which all our lives depend.' Part of that knowledge has to include facts about animal agriculture so that people can better understand the consequence of their purchases and lifestyles.

Three years after writing two grumpy letters to RGB Kew complaining about the lack of vegan options and how Kew should be at the head of the queue within the vegan movement, we were pleased to see three good vegan options on the lunch menu at their refurbished Pavilion restaurant.[16] But why stop there? The logical step is for both RGB Kew and the RHS to serve only plant-based food at their establishments and events to reinforce their respective mission statements and dedication to the stewardship of the immediate and wider environment. If they think people won't visit if they can't consume animal products, then clearly the gardens aren't interesting enough.

With the current scale of species extinction (over 137 plant, insect and animal species lost each day due to deforestation caused by animal agriculture),[17] it's hard to imagine a more alarming statistic and what else might be needed to make people address the issue of meat consumption. Thankfully, outside the comfort zone of the garden, some people are speaking up. Environmental activists inspired by Extinction Rebellion commit acts of peaceful civil disobedience blocking roads, bridges and buildings to raise awareness of the impending climate disaster. Vegans (who have watched *Cowspiracy*) have been banging on about it for years, and one of Kew's most celebrated and respected neighbours, David Attenborough, represented the people of the world at the 2018 UN climate change summit in Poland. His address to the delegates should sound alarm bells from the quiet corridors of the herbarium to the top of the recently restored pagoda: 'If we don't take action, the collapse of our civilizations and the extinction of much of the natural world is on the horizon.'

Of course, it's great to learn from the RHS how to grow food locally, encourage wildlife, build a green roof, deal with the effects of flooding or make a rain garden. It's also good to know that RGB Kew has the world's most diverse global herbarium and fungarium and that, together with its Sussex botanic garden, Wakehurst Place, its wild plant seed bank is being curated and conserved for the advancement of science and the benefit of the human race. But until one or the other (ideally both) start advocating for a shift away from the consumption of animals to the consumption of plants, and educating their members about the reasons why, their *raison d'être* has to be questioned. By ignoring the elephant in the room they are letting down not only their members and the planet that sustains them, but also future generations of all sentient beings left to reap nothing but the shriveled, ghostly husks of seeds we failed to sow.

EDUCATION

Talking of elephants, in 2007 Christine and I were fortunate to visit India and Nepal and get a taste of the country my mother's side of the family knew back in the 1930s and 1940s. At the time we were vegetarian, eight years away from being vegan, and travelling in a country where vegetarianism was the norm, we were well provided for at our various lodgings.

One of the highlights towards the end of the trip was a jungle safari at Chitwan National Park where part of the package included an elephant ride deep into the jungle to find the infamous one-horned rhino often referred to as Dürer's rhinoceros. It was almost as enjoyable as we'd anticipated (we had the good fortune to see three rhinos in total), but there was something very disturbing about the experience as well. Our Nepali guide 'driving' the elephant (known as a mahout) used a metal spike to steer it, often poking its ear, hitting its head or

'IF YOUR PLAN IS FOR ONE YEAR PLANT RICE. IF YOUR PLAN IS FOR TEN YEARS PLANT TREES. IF YOUR PLAN IS FOR ONE HUNDRED YEARS EDUCATE CHILDREN.'

CONFUCIUS

jabbing the spike hard into the base of its skull if it took a wrong turn or stopped for a cheeky munch on passing foliage. On returning to the camp we realized our naivety. With chains on their feet and the young separated from their mothers in their enclosures, these noble creatures were nothing more than circus animals. We could only guess at the ordeal they were put through for their spirits to be broken in order for humans to ride them.[18] The following day, as we bathed with one of the elephants in the East Rapti River, we knew we had funded animal exploitation.[19] Since becoming vegan we have learned more about the methods used to train wild animals, and the shame of having taken part in such exploitation doesn't sit well at all.

The reason I mention it here is to give an example of how someone who respects animals can still be blind to the obvious. We've never wanted to visit Sea World to see dolphins and killer whales because it's so blatantly cruel to keep such creatures confined in a glorified fish tank, so why did we think that riding elephants into a jungle was okay? We had the best of intentions but, ultimately, we were naive and completely oblivious to their misery. Education, therefore, is the only way we can make things better for animals, and this is why it's important to speak up for their rights.

Our tolerance of what we do to animals for food, experimentation, clothing or entertainment is generally referred to as 'speciesism': that is, these same practices would be judged as criminal if performed on a member of our own species. It's very convenient for us that these beings either walk on all fours or have fur, scales, feathers, horns, beaks or gills; their physical attributes make it difficult for us to relate to them. However, we know they are beings with similar traits and desires: the desire to live a life of freedom and the right to live a life free from the unnecessary torture or suffering caused by humans. Recognizing speciesism and ending this form of discrimination is a giant step towards a peaceful, compassionate and sustainable world, but some people still see this as a threat to their personal choices.[20] In many minds, the word 'advocate' is not a million miles from the word 'militant' or 'extremist'. It's often the way where injustices are being fought and societal norms are being challenged. Many of those advocating for a vegan way of life may prefer to use the word 'educator'.

NORMALIZED VIOLENCE

Educating people who have lived all their lives in a system that profits from oppression is challenging to say the least. While children have the capacity to see the world with more clarity and tend not to muddy logic with their selfish preferences, adults are more difficult. Many are indifferent and resistant to the suggestion of seeing the world from another perspective. Underpinning this resistance to end the persecution of animals is a lifetime of conditioning:

> Normalized violence is the underlying foundation which our society has agreed on so that the commodification of farm animals for food would be considered 'normal.' However, we have created the most egregious act of violence against our own humanity by accepting violence as a necessary ingredient, not only on our respective plates, but we have accepted the notion that it is 'normal' and 'right' to pass the same bloody traditions down to our own children. The family system is wrought through and through with rituals around food that cause family structures to be dependent on violence as a way of life. When a family member rejects animal foods and adopts a plant-based diet, often times other family members will feel betrayed and rejected. That is how strong the bonds that we have invisibly agreed upon around the suffering and death of innocence are.[21]

Those who do understand the seriousness of the situation but are unwilling to commit to being vegan will often say things like 'it's a question of balance', 'everything in moderation', 'a flexitarian approach is best', 'I make sure my meat is ethically sourced'. None of these are arguments against veganism; they are merely statements that suggest an ambivalent attitude to the main issues of animal cruelty, environment and health.

Any argument that keeps the exploitation of animals on the agenda, no matter how good the level of welfare, will always end with their unnecessary slaughter and set a precedent that animal abuse is acceptable. It might seem unreasonable to suggest that the elimination of animal agriculture is the only reasonable course of action, but considering that the sheer numbers of animals killed each year are beyond anyone's imagination, many would argue this is the least we can do to atone for the suffering they've endured for so long.

Pushback from the meat and dairy industries is inevitable. Just as the tobacco industry used to refute the claims that smoking is bad for you, there's too much money to lose for them not to fight back. This will take the form of ignoring the overwhelming evidence that they are cruel, bad for the climate, and bad for environmental and personal health. They will pick up on one thread in an argument against veganism that has an element of truth in it and then use it as a smokescreen to validate the choice of non-vegans.[22] They may even publish their own reports funded by themselves or others with a vested interest in profiting from such industries.

WHAT WILL HAPPEN TO THE ANIMALS?

A frequent question when discussing veganism is 'what will become of the animals in a vegan world?' The bucolic idyll is such a potent notion for our species. The images of sheep and cattle grazing in fields and ponies roaming the wilds of Dartmoor and Exmoor in peace, harmony and happiness ignore the reality of what is nothing less than systematic oppression and violence on an unimaginable scale.[23]

While the number of animals would fall, it would not happen overnight. When animals in industrialized farms are consigned to the history books, they won't necessarily have to disappear altogether. Rewilding projects such as Knepp Farm (see page 101) show that animals may well have a role to play in reforesting the world and its management. Aside from private estates and sanctuaries where animals could be seen in traditional fields, the countryside may well change, with more trees and scrub, but it's a small price to pay for the increased diversity that comes with it.

CONSUMER POWER

For some reason it's easier to cope with backlash from aggressive carnists then the apathetic attitudes of people you know. One of the weakest arguments I've heard – and I've heard it a number of times over the last few years – is 'well, they drop bombs on Syria.' It makes no sense whatsoever, but I think what they're trying to say is that there's not much we can do about it. While world conflict might be beyond our control, we as consumers have enormous power collectively. Simply choosing plant-based options over animal products sends a strong message to the food industry, which is already changing as the vegan snowball gathers pace. By choosing a plant-based diet we could mitigate climate change and deforestation, species extinction, ocean depletion, pollution and put an end to world hunger. Imagine being able to make real and positive changes like these simply by changing your food choices. Imagine also being able to live a far healthier life and eliminate unnecessary harm to billions of innocent lives. If people are inherently good, then it's not unreasonable to think that they will buy into this.

Since becoming vegan I've seen the very best of what the human race has to offer but I've also seen the very worst. I've struggled with it at times, partly because of the sheer scale of the horror, partly because I'm impatient where injustice is concerned, and also because I've underestimated just how conditioned we've become to the all-powerful propaganda of the meat and dairy industries. That power, though, is only given to them by us, the consumer. The truth is that we have the power to change things for the better by not buying these products and choosing plants.

The good news is that it's happening. In 2014 there were around 150,000 vegans in the UK. A survey carried out by the Vegan Society in 2018 suggests that this figure had quadrupled to approximately 600,000.[24] More and more vegan restaurants are opening, and there are vegan substitutes for pretty much everything you think you might miss. I can honestly say that I've never eaten so well as I have done in the last four years and have felt better for it too.

NOW YOU KNOW THE FACTS, WHAT ARE YOU GOING TO DO?

We know that we can survive without eating meat. We know that a plant-based diet will put an end to the unnecessary cruelty to trillions of animals each year. We know that it will make us healthier and help to offset some of the worst environmental catastrophes currently in play. We know that a vegan approach to life could be the best way of preserving life as we know it for our children and future generations. So, what's it going to be?

1 I didn't know the facts but now I'm informed I'm going to make some simple changes and make this a better world.
2 I didn't know the facts and in spite of being informed I still don't really care about animals and the ecological/health consequences of eating them.

There are many people more inclined to reduce their meat consumption rather than give it up altogether. Certainly, this softer 'flexitarian' approach is better than nothing. The flexitarian sees meat as a treat, and this is much more appealing to those who know the health and ecological impacts of animal agriculture but can't imagine giving up meat altogether. Increasing the quality and making sure that meat is produced more locally are also notions that are less challenging to passionate meat-eaters who feel they have a right to eat what they want. I have mixed emotions about this. While anything that brings down the demand for meat is a good thing, there are problems. First, the price of meat, already heavily subsidized, would skyrocket.[25] Only the well-off would be able to afford to consume it on a regular basis. Those in lower income groups would have to look at it as a rare pleasure. Second, it reinforces the notion that high welfare meat is morally acceptable, which, to an ethical vegan, it isn't. Imagine someone saying to activists in any equal-rights movement that their demands will be met but only on certain days of the week. Imagine someone saying that they refrain from modern day slavery over the weekend.

CONCLUSION

In thinking about how to conclude this book, Ted Hylands came to mind. Ted taught me the tricks of the landscaping trade when I first expressed an interest in gardening. Here's an article I wrote about him for the *Independent* magazine in 2004 to celebrate the 60th anniversary of D-Day:

> Ted's coming to lunch tomorrow at the allotment and we're keen to put on a good spread. It's his first visit to the plot, which is looking reasonably good at the moment largely

due to the flowers rather than the vegetables. The herb bed in particular has already seen *Allium* 'Purple Sensation', borage and eschscholzias provide the first flush of colour. Pot marigolds that survived the winter have also given the plot a lift, blanketing one corner a clear orange. Now drumstick heads of *Allium sphaerocephalon* with that edible green to wine-purple transition, providing edge-of-your-seat anticipation, will combine with self-seeders such as fennel and *Verbena bonariensis* to provide one enormous feeding station for nectar-hungry insects.

Incredibly, I've known Ted for 25 years but can't remember talking to him about growing vegetables. He taught me just about everything I know about hard landscaping and kept me working during a difficult period when I finally became self-employed. At 78, and with 50 years landscaping behind him, he's starting to take things easier these days. His Ford Thames van with a Perkins diesel engine, a familiar sight around Leatherhead, Surrey, is symbolic of his reliability and facility to keep going at a steady pace (though there were plenty of winter days when I cursed the fact it had no heating). His collection of mowers and other garden machinery and his ability to fix or customize them are legendary. My favourite, a behemoth rotovator like some ageing dinosaur that would have to be cranked up into spluttering life, would carve up a plot in minutes if you were strong enough to reign it round corners.

His disciplined manner belies his patience, trust and ability to let things go when I made mistakes with levels or took infernally long to put down the first course of a brick wall. Clumsy habits with tools were ironed out and techniques for cutting and dressing stone without power tools saved numerous trips to the hire shop.

I did think him slightly mad using hedge trimmers on roses and restoring a laurel hedge by cutting the whole thing almost to the ground. Both responded with grateful vigour. I winced at the tacky choice (on the part of the client) of fibreglass columns for a folly that had to be filled with concrete and was secretly pleased when a large cedar fell on it during the Great Storm of 1987. We shared the disheartening experience of having to rebuild it, again with more fibreglass, despite having sourced reclaimed-stone columns at a bargain price.

At my first show garden at Hampton Court, ten years ago, Ted was there, reliable as ever, ready to build, plant and give the moral support that's vital to see you through such an event. Celebrating the then 50th anniversary of the D-Day landings, I expressed my relief that he'd been too young to be involved, joking that he may not have been around to help me if he had. His response that, aged 18, he'd served with the Royal Marines landing at Sword Beach came as a shock. I had known him for

> 14 years and this was the first time he'd mentioned it. It's not uncommon for those who've experienced the reality of war to keep it to themselves and Ted, matter-of-fact as ever, brushes it aside as he would leaves off a path. No stories of valour, hardship or sorrow. 'It was a job that needed to be done and I was just fortunate to come back, that's all.'
>
> So, together with his wife, Iris, some family and friends, we'll celebrate his return at our plot, a suitable venue in many ways. Apart from the obvious 'Dig for Victory' connotations of providing food during the two world wars, plans for the Normandy landings were drawn up by General Eisenhower just a stone's throw away in Bushy Park. To cap it all, having planted our first earlies three months ago now, there's an outside chance I'll be able to serve up a bowl of new potatoes, the first of the season, garnished with fresh mint and a generous knob of butter. What variety? Blow the sentiment, 'Winston', of course.

In 2011, we had the privilege of taking Ted back to Sword Beach in Normandy for the first time since the war. Being with war veterans and visiting museums, monuments and the beaches themselves made for a sobering and poignant few days. The significance of what was achieved by those involved, not to mention the bravery, is beyond words. Quite simply, we are forever in their debt for our freedom.

Later, on the 75th anniversary of D-Day, Ted received nothing less than royal treatment (including lunch with the Queen) with fellow veterans during a week of celebrations culminating in a trip aboard MV Boudicca to relive the journey they and their comrades made three-quarters of a century ago.

While it would be foolish to draw any parallels between the environmental crisis we find ourselves in and the evils of fascism, there is a sense that we've reached a point where it would be equally foolish to ignore the damage we have caused to the natural world. But just as the youth of the nations involved in the Second World War secured our freedom, today's youth understand that their future is once again at stake but in a different way. Some, like Greta Thunberg and thousands of schoolchildren around the world, are prepared to stand up and fight for it.

The problem is that this threat isn't as obvious as a Nazi flag, the Blitz or fighting anyone on the beaches; it's our indifference to the consequences of the way we live, what we consume and how we dispose of these things. The problem is us. We are a remarkable species, one of many remarkable species that inhabit a remarkable world, but we are flawed. If we could just get over the misguided notion that the universe centres around us, we might be able to leave this world in a state that we would have liked to find it. With a big enough shift in consciousness we might even be able to improve it.

I hope that this book will be out of date by the time it's published, and that instead of 600,000 vegans there will be two million. I hope that those who think veganism is trying to take something from them will soon understand that it is a gift. I hope that those who speak up for the voiceless will not be made to feel guilty or timid for protecting the defenceless by carnists looking to deflect the guilt that they feel in perpetuating animal cruelty for no good reason. I hope that those who understand the true meaning of being vegan will become active in the struggle for animal rights to help speed up the transition to a vegan world, so that those carrying the burden of animal advocacy can all get back to enjoying this bonkers world and everything it has to offer.

As gardeners I know we have the sensitivity and ability to make the changes necessary to sustain a healthy world. But we also need to create a tsunami of awareness and educate others about what we do, how life-affirming it can be, and how some simple changes to the way we live will have long-lasting benefits to future generations. Our future is in our hands; it's all about the soil, it's all about our oceans, it's all about compassion and, best of all from a gardener's point of view at least, it's all about plants.

If you've read this far, thank you. Even if you still have doubts about everything I've written or questions about some of the statistics, please leave animals off your plate and give plants a chance. Going vegan has been the most life-affirming experience and reinforces the notion that 'the meaning of life is to give life meaning.' Like many others the only regret I have about going vegan is not doing it sooner. The values I was taught as a child and at school – to be kind and treat everyone as an equal – have been reinforced a thousand times by living a vegan ethic. To discover that I am working in an industry and among people who might just be able to secure a future for our children's children has been a revelation to me, and naturally I want to share it. You can ask me to shut up, but I'm afraid I won't. I've seen too much. I've seen the truth.

Gardeners have a natural predilection for all life. We understand the power and the magic of nature and what it provides. If we know and enjoy how gardens feed and heal us, then isn't it time for us to give something back? It's time to ask not what the garden can do for you, but what you can do for the garden.

THE GARDEN OF EDEN IS BROKEN. THE GARDEN OF VEGAN CAN FIX IT.

AL MA'ARRI (AD 973–1057),
'I NO LONGER STEAL FROM NATURE'

YOU ARE DISEASED IN UNDERSTANDING
AND RELIGION.
COME TO ME THAT YOU MAY HEAR
SOMETHING OF SOUND TRUTH.
DO NOT UNJUSTLY EAT FISH THE WATER
HAS GIVEN UP,
AND DO NOT DESIRE AS FOOD THE FLESH
OF SLAUGHTERED ANIMALS,
OR THE WHITE MILK OF MOTHERS WHO
INTENDED ITS PURE DRAUGHT FOR THEIR
YOUNG, NOT NOBLE LADIES.
AND DO NOT GRIEVE THE UNSUSPECTING
BIRDS BY TAKING EGGS; FOR INJUSTICE
IS THE WORST OF CRIMES.
AND SPARE THE HONEY WHICH THE BEES
GET INDUSTRIOUSLY FROM THE FLOWERS
OF FRAGRANT PLANTS;
FOR THEY DID NOT STORE IT THAT IT MIGHT
BELONG TO OTHERS,
NOR DID THEY GATHER IT FOR BOUNTY
AND GIFTS.
I WASHED MY HANDS OF ALL THIS; AND
WISH THAT I
PERCEIVED MY WAY BEFORE MY HAIR
WENT GREY.

NOTES

INTRODUCTION

1 The word 'holocaust' literally means 'burnt offering' (Leviticus 1:1–10). It is often used by animal rights activists, along with words such as slavery, rape and murder, to provoke an emotional response. I've used the word myself but have come to understand that it can cause offence to anyone, including vegans, who has suffered such a misfortune.

1. SEEDS

1 Later as a pre-vegan adult I again had a similar experience while fishing the Little Great Ouse with my father from his back garden. It put an end to my fishing days for good.
2 Follow @Seaspiracy and @LetFishLive on Instagram for the latest facts and figures.
3 T. E. Higham et al., 'Angling-induced injuries have a negative impact on suction feeding performance and hydrodynamics in marine shiner perch, *Cymatogaster aggregata*', *Journal of Experimental Biology* (October 2018).
4 My brother has been vegan since 2016.
5 For all the machismo associated with meat, it's ironic that meat consumption is a known cause of prostate cancer and erectile dysfunction.
6 'Carnist' is a term coined by Melanie Joy in her book, *Why We Love Dogs, Eat Pigs and Wear Cows: An Introduction to Carnism* (2011).
7 Needless to say, the RSPCA officer took a dim view of this, and by the time the fox was put into the vehicle, that brief moment of trust in its eyes was gone. Thankfully, the fox made a good recovery and was released a week later.
8 The definition of the word 'humane' is 'tenderness, compassion and sympathy for people and animals'.
9 It's only since being vegan that I've learned that Olympic champion Carl Lewis adopted a plant-based diet during his medal-winning years.
10 Information about efficient, low-cost alternatives to culling can be found on Animal Aid's website: www.animalaid.org.uk – 'Alternatives to Culling'.
11 See 'The myth of the evil grey squirrel' on www.urbansquirrels.co.uk.
12 The grey squirrel is often derided as a non-native, but existing populations of red squirrel can't be accurately described as native either.
13 Animals killed in the Royal Parks include deer, rabbits, squirrels, foxes, crows, geese, jays, magpies, parakeets and pigeons. Freedom of Information Act requests by a national animal rights organization, Animal Aid, revealed that more than 11,000 animals were culled in the Royal Parks between January 2013 and January 2017 (inclusive).
14 See 'Royal Parks cull leaflet' on www.urbansquirrels.co.uk.
15 Animal Aid, '"Royal Slaughter": Extent of Royal Parks' wildlife culling revealed', 7 October 2017, www.animalaid.org.uk.
16 Humans would be unable to survive on earth without these animals, which provide vital services in pollination, decomposition, seed dispersal and pest control.
17 It pains me to admit it, but a temporary setback occurred when we became so overrun with slugs during a particularly wet spring that I resorted to feeding them to chickens. This is documented in my first book, *Our Plot*, along with a period when we went back to eating fish.
18 Having spent most of my youth trying to fulfil an Olympic ambition in track and field athletics, a combination of aches and pains and gardening-related back injuries forced me to hang up my training shoes when I was forty. Almost twenty years later with the aches subsiding I tentatively tried jogging. Every time I'd tried this before had ended with a trip to the osteopath, but this time it felt different. After several overenthusiastic sessions stirred up an old knee injury. I may have to accept that my running days might be over, but I like the fact that I'm even entertaining the notion of getting fit (in the old sense of the word) again.
19 K. Stallwood, *Growl: Life Lessons, Hard Truths, and Bold Strategies from an Animal Advocate* (2014).
20 Joy, *Why We Love Dogs, Eat Pigs and Wear Cows*.
21 Worldwide, in 2018, there were five fatal shark attacks: 'Yearly worldwide shark attack summary' from Florida Museum, 2018, www.floridamuseum.ufl.edu. Humans kill approximately 63–273 million sharks per year: Boris Worm et al., 'Global catches, exploitation rates, and rebuilding options for sharks', *Marine Policy*, 40 (2013), pp. 194–204.
22 You don't give up anything by being vegan; you just stop taking what isn't yours.
23 Visit Earthling Ed's website (www.earthlinged.com) and watch his video series '30 Days, 30 Excuses'.
24 From Andrew Kirchner's website, www.kirschnerskorner.com.
25 Cleve Backster, *Primary Perception: Biocommunication with Plants, Living Foods and Human Cells* (2003).
26 While physiologically we are herbivores or frugivores, our history of eating animals leads one to assume that we are omnivores. If you need convincing please watch lectures by Dr Milton

Mills on YouTube. Even if we were omnivores, it doesn't mean that we are obliged to eat meat; it simply means that we are capable of eating meat. It is not necessary to eat meat to survive.

27 Worldwide, at least 50 per cent of the grain grown is fed to livestock (*Cowspiracy*).

28 J.L.P., 'Meat and greens: How bad for the planet is eating meat?, *Economist* (31 December 2013), www.economist.com. According to www.cowspiracy.com, animal agriculture is responsible for up to 91 per cent of Amazon rainforest destruction.

29 Processed vegan products are not much cheaper and can be more expensive than processed products made from animals due to the fact that much of the meat and dairy industries is subsidized. As the demand for vegan food increases, this will reverse. Wholegrains, fruit and vegetables are significantly cheaper than meat.

30 McDonald's Corporation *v.* Steel & Morris (1997), known as the McLibel case, involved activists accusing the corporation of encouraging the mistreatment of animals and workers, litter and the destruction of rainforests.

31 Ipsos Mori survey commissioned by the Vegan Society in 2018 and 'The Food & You Survey' by the Food Standards Agency in 2017.

32 Peter Egan Facebook page, 8 November 2018. (I'm a little jealous of Peter's surname. Just imagine: CleVegan).

2. VEGANICS: VEGAN ORGANIC GARDENING

1 From a vegan point of view, I now see the Green Man *eating* leaves, not disgorging them, a sign of a plant-based future?

2 Views expressed by panellists on *Gardeners' Question Time* have improved a little over the years, but even organic pundits still manage to bring you up with a start with their intolerance, blinkered views and blitzkrieg tactics when insects and animals (labelled as pests) compete for the food they are trying to grow.

3 J. Hall and I. Tolhurst, *Growing Green: Organic Techniques for a Sustainable Future* (Vegan Organic Network, 2006, rev. edn 2010).

4 Ibid.

5 I've often wondered about what sustains rainforests, which one might imagine to be quite demanding in terms of nutrients. One might assume that the animals in the forest provide the necessary nutrients, but the reality is that plants (leaf litter and dead trees) provide the bulk of a rainforest's needs.

6 We should try to understand that this is not the insects' fault and that many such infestations and diseases have been caused by human intervention where plants have been shipped around the world.

7 Even if it was a wasps' nest we would leave it well alone. Their presence is transitory (one year) and they are also important predators and pollinators.

8 Truth or Drought, 'Myth: Beekeeping is needed to conserve pollinator populations', www.truthordrought.com.

9 Ibid.

10 G. Pearson, 'You're worrying about the wrong bees', *Wired* (29 April 2015), www.wired.com.

11 There are over 275 bee species in the UK made up of six families: www.lbka.org.uk.

12 Unscrupulous tenants who have a reputation for stealing them to eat eventually put me off the idea.

13 Urine contains useful levels of nitrogen, phosphorus and potassium with an N-P-K ratio of 11-1-2.5. It's a good activator that can be added neat to carbon-rich compost heaps or watered down with 8–10 parts of water to make a liquid feed.

14 Vegan Organic Network article by Meghan Kelly, 'Fertilising with human urine' (14 October 2018), www.goveganic.net.

15 Strictly speaking, as it is derived from a human animal, this is not plant-based, but assuming you're using your own pee and poo and not from a non-vegan (and certainly not from someone who is being exploited in any way), my conscience would be clear if I ever get round to experimenting with humanure.

16 See www.envirolet.com.

17 *The Humanure Handbook: Shit in a Nutshell* is available at www.humanurehandbook.com.

3. HEALING GARDENS

1 Taken literally, the title 'Healing Gardens' could be construed as misleading. It's unlikely that a garden will alter the eventual outcome of diseases such as Alzheimer's, though it can offer relief from psychological stress associated with the condition. Gardens do, however, provide a useful resource for attaining therapeutic goals related to motor and cognitive skills, not to mention psychological well-being, which might be part of a holistic treatment programme to treat a whole gamut of medical conditions.

2 Catharine Paddock, PhD, 'Soil bacteria work in similar way to antidepressants', *Medical News Today* (2 April 2007).

3 There were two exceptions: once when I used a beer trap for slugs in the first year and another when some slugs were fed to a neighbour's chickens. The guilt still haunts me today.

4 Apologies for not being able to come up with a vegan version of this phrase.

4. THE ENVIRONMENT

1 It's argued that the Anthropocene started in the late eighteenth century as concentrations of carbon dioxide and methane have been found trapped in polar ice of this period – the same time as the invention of the steam engine: P. J. Crutzen, 'Geology of mankind', *Nature* (2002). See also www.anthropocene.info.

2 Compiled by 145 expert authors from fifty countries over three years with inputs from another 310 contributing authors, the IPBES Global Assessment is the most comprehensive report of its kind.
3 Only two British newspapers, the *Guardian* and the *Independent*, made room for the biodiversity story on their front page.
4 Damian Carrington, 'Humans just 0.01% of all life but have destroyed 83% of wild animals – study', *Guardian* (21 May 2019).
5 The earth's limit for resources (minerals, fossil fuels, metal ores, contruction minerals, industrial minerals and biomass) is approximately 50 billion tonnes per year. At the time of writing we use approximately 70 billion tonnes, and taking into account predicted growth rates, it will be approximately 180 billion tonnes by 2050. Even if we did everything in our power (innovation, efficiency, hefty carbon taxes), we can only hope to reduce this to 95 billion tonnes, almost double what the planet can sustainably provide. See George Monbiot, 'The earth is in a death spiral: It will take radical action to save us', *Guardian* (14 November 2018).
6 J. Poore and T. Nemecek, 'Reducing food's environmental impacts through producers and consumers', *Science* (June 2018)
7 See FAO, 'Livestock's long shadow: Environmental issues and options', Food and Agriculture Organization of the United Nations, Rome (2006), www.fao.org.
8 Poore and Nemecek, 'Reducing food's environmental impacts'.
9 Eutrophication is the leaching of nutrients (often from agricultural land) into water bodies that can cause algal blooms and prevent oxygen from getting into the water, creating dead zones where fish and other organisms can't survive.
10 P. J. Gerber et al., 'Tackling climate change through livestock: A global assessment of emissions and mitigation opportunities', Food and Agriculture Organization of the United Nations, Rome (2013), www.fao.org.
11 P. D. Smith et al., 'Agriculture', in B. Metz et al., eds, *Climate Change 2007: Mitigation of Climate Change; Contribution of Working Group III to the Fourth Assessment Report of the Intergovernmental Panel on Climate Change* (Cambridge University Press, 2007), https://www.ipcc.ch/report/ar4/wg3/.
12 G. Myhre et al., 'Anthropogenic and natural radiative forcing', in T. F. Stocker et al., eds, *Climate Change 2013: The Physical Science Basis; Contribution of Working Group I to the Fifth Assessment Report of the Intergovernmental Panel on Climate Change* (Cambridge University Press, 2013), https://www.ipcc.ch/report/ar5/wg1/.
13 D. Tilman and M. Clark, 'Global diets link environmental sustainability and human health', *Nature* (2014).
14 The suggestion is that constant grazing stimulates plants to grow and encourages root growth, thereby storing more carbon.
15 R. McKie, 'We must change food production to save the world, says leaked report: Cutting carbon from transport and energy "not enough" IPCC finds', *Guardian* (4 August 2019).
16 FAO, 'The state of food and agriculture: Livestock in the balance', Food and Agriculture Organization of the United Nations, Rome (2009), www.fao.org.
17 Tara Garnett et al., 'Grazed and confused', Food Climate Research Network, University of Oxford (2017), www.fcrn.org.uk.
18 FAO, 'Livestock's long shadow'.
19 Helen Harwatt and Matthew Hayek, 'Eating away at climate change with negative emissions: Repurposing UK agricultural land to meet climate goals', Harvard University, Cambridge, Mass. (11 April 2019), https://animal law.harvard.edu.
20 'For every 100 food calories of edible crops fed to livestock, we get back just 17 calories in the form of meat and dairy: an 83 per cent loss.' See 'This archaic method of food farming has failed', www.ciwf.org.uk
21 Charlotte McDonald, 'How many earths do we need?', BBC News (16 June 2015), www.bbc.co.uk.
22 While the biggest cost for grass-fed systems is land, intensive systems are not exactly cheap either. Stored forage or feed crops need to be grown, harvested, transported, stored and fed, and this requires a not insubstantial investment by way of fuel, equipment, maintenance and labour. Subsidies obviously play a big part in keeping these costs down and attractive to the consumer.
23 G. Henderson et al., *Greenhouse Gas Removal* (Royal Society and Royal Academy of Engineering, 2018), www.royalsociety.org.
24 Harwatt and Hayek, 'Eating away at climate change with negative emissions'.
25 David D. Briske et al., 'Commentary: A critical assessment of the policy endorsement for holistic management', *Agricultural Systems* vol. 125 (March 2014, pp. 50 –53. See also George Monbiot, 'Eat more meat and save the world. The latest implausible farming miracle', *Guardian* (4 August 2014)
26 Email from Jenny Hall to the author.
27 In the UK, grasslands are the main user of nitrogen, with 425 kilotonnes applied in 2015. Permanent grassland used 73 per cent of this amount. See Table A 3.3.8, Areas of UK crops and quantities of fertiliser applied for 2016: https://uk-air.defra.gov.uk/assets/documents/reports/cat07/1804191055_ukghgi-90-16_Annexes_Issue1.1_UNFCCC.pdf.
28 William J. Ripple et al., 'Are we eating the world's megafauna to extinction?', *Conservation Letters*, vol. 12 (February 2019), Society for Conservation Biology.
29 UNFAO, 'Livestock primary: Producing animals/slaughtered 2017' (2019), www.fao.org/faostat/en/#data/QL

30 Bibi van der Zee, 'Why factory farming is not just cruel – but also a threat to all life on the planet', interview with Philip Lymberty, head of Compassion in World Farming and author of *Farmageddon* and *Deadzone*, in the *Guardian* (4 October 2017), www.theguardian.com.
31 A. Shepon et al., 'The opportunity cost of animal based diets exceeds all food losses', *Proceedings of the National Academy of Sciences of the United States of America*, 115 (2018), pp. 3804–9.
32 What's not necessarily taken into consideration is that it's not obligatory to consume soya as a vegan. All fruit, vegetables, seeds, grains, nuts and pulses contain protein, and it's relatively easy to get all the protein, vitamins and minerals you need through a balanced wholefood diet.
33 B. Machovina and K. J. Feeley, 'Meat consumption as a key impact on tropical nature: A response to Laurance et al.', *Trends in Ecology and Evolution* (29 August 2014).
34 Luis Alonso Lugo, 'US, Brazil hold talks on trade, Amazon protection', *Washington Post* (13 September 2019).
35 C. A. McAlpine et al., 'Biodiversity conservation and vegetation clearing in Queensland: Principles and thresholds', *Rangeland Journal* (15 June 2002). C. A. McAlpine et al., 'Increasing world consumption of beef as a driver of regional and global change: A call for policy action based on evidence from Queensland (Australia), Colombia and Brazil', *Global Environmental Change* (2009). WWF, 'Australian animals lost to bulldozers in Queensland 2013–15', www.wwf.org.au.
36 S. L. Maxwell et al., 'Degradation and forgone removals increase the carbon impact of intact forest loss by 626%', *Science Advances* (30 October 2019). See also G. Readfearn, 'Climate emissions from tropical forest damage "underestimated by a factor of six"', *Guardian* (31 October 2019).
37 Ripple et al., 'Are we eating the world's megafauna to extinction?'
38 'Nature's Dangerous Decline "Unprecedented"; Species Extinction Rates "Accelerating"', IPBES Global Assessment Report on Biodiversity and Ecosystem Services (2019), www.ipbes.net.
39 Bacteria alone account for 13 per cent of all life on the planet. Yinon M. Bar-On, Rob Phillips and Ron Milo, 'The biomass distribution on earth', *Proceedings of the National Academy of Sciences* (2018).
40 FAO, 'Soils are endangered, but the degradation can be rolled back' (4 December 2015), www.fao.org.
41 R. J. Zomer et al., 'Global sequestration potential of increased organic carbon in cropland soils' (14 November 2017), www.nature.com.
42 A. L. Sansom, 'Upland vegetation management: The impacts of overstocking', *Science Direct*, 39 (1999). See also George Monbiot, 'Sheepwrecked', *Spectator* (30 May 2013), www.monbiot.com.
43 DEFRA memorandum submitted to Parliament, 'Farming in the Uplands' (2010), www.parliament.uk.
44 The obvious concern here from a vegan point of view is that it becomes a distraction from the vegan philosophy, which is based not on welfare but on avoiding the deliberate killing or exploiting of animals.
45 Arthur George Tansley (ecologist) suggests that being overrun with wild animals is unlikely to happen (see p.151).
46 'Rape-rack' is the dairy industry's term used for a device that restrains cows while they are forcibly inseminated.
47 Boris Worm et al., 'Impacts of biodiversity loss on ocean ecosystem services', *Science*, vol.314 (3 November 2006).
48 See Conservation International, 'About blue carbon' (2019), https://www.thebluecarboninitiative.org/about-blue-carbon.
49 University of Maryland, 'Coastal wetlands excel at storing carbon: New analysis supports mangrove forests, tidal marshes and seagrass meadows as effective climate buffers', *ScienceDaily* (1 February 2017). See also 'Mangrove forests are one of the world's most threatened tropical ecosystems', www.wwf.panda.org.
50 T. B. Atwood et al., 'Predators help protect carbon stocks in blue carbon ecosystems', *Nature Climate Change*, 5 (2015), pp. 1038–45.
51 There are almost four hundred dead zones worldwide, affecting over 70,000 square miles. See J. Howard, 'Dead zones, explained: Few marine organisms can survive the toxic low-oxygen conditions of dead zones. Here's how our agricultural practices make them worse', *National Geographic* (31 July 2019).
52 On the accidental killing of dolphins and whales, see 'Bycatch is the biggest killer of whales: Time for the world to tackle the threat' (20 October 2016), www.wwf.panda.org.
53 Trout are often stocked so densely that there are between thirteen and twenty-seven 30-centimetre-long fish per bathtub. See P. Lybery, 'In too deep: Why fish farming needs urgent welfare reform' (2002), ciwf.org.uk.
54 It's estimated that seafood eaters ingest up to 11,000 tiny pieces of plastic every year. See S. Knapton, 'Seafood eaters ingest up to 11,000 tiny pieces of plastic every year, study shows', *Telegraph* (24 January 2017).
55 The Waitrose Food and Drink Report 2018–19 showed that most consumers changed how they use plastic after viewing *Blue Planet*: 44 per cent = drastic change; 44 per cent = somewhat changed; 12 per cent = didn't change; see www.waitrose.com.
56 640,000 tonnes of fishing gear are discarded in the ocean every year. See 'Ghosts Beneath the Waves', *World Animal Protection* (2018) and Laura Parker, 'The great Pacific garbage patch isn't what you think it is. It's not all bottles and straws – the patch is mostly abandoned fishing gear'. *National Geographic* (22 March 2018).

57 See www.cowspiracy.com/facts.
58 Peter Scarborough et al., 'Dietary greenhouse gas emissions of meat-eaters, fish-eaters, vegetarians and vegans in the UK', *Climatic Change*, vol. 125 (July 2014), pp. 179–92.
59 Look no further than TUCO's greenhouse gas footprint calculator: www.tuco.ac.uk/ghgcalculator/index.html.

5. HEALTH AND HUMANITY

1 'The Dietary Reference Values for protein are based on estimates of need. For adults, an average requirement of 0.6 g of protein per kilogramme bodyweight per day is estimated. The Reference Nutrient Intake (RNI) is set at 0.75 g of protein per kilogramme bodyweight per day in adults. This equates to approximately 56 g/day and 45 g/day for men and women aged 19–50 years respectively. There is an extra requirement for growth in infants and children and for pregnant and breastfeeding women.' See British Nutrition Foundation, www.nutrition.org.uk.
2 'British Dietetic Association confirms well-planned vegan diets can support healthy living in people of all ages', *BDA News* (7 August 2017).
3 K. L. Tucker et al., 'Plasma Vitamin B-12 concentrations relate to intake source in the Framingham Offspring Study', *American Journal of Clinical Nutrition* (February 2000), pp. 514–22.
4 250 mcg per day or 2,500 mcg per week: see Dr M. Greger, 'Vitamin B12: How much, how often' (30 August 2011), www.nutritionfacts.org.
5 PCBs are highly toxic industrial compounds that can cause serious developmental and neurological problems in foetuses, babies and children. Banned in the USA in 1977, PCBs are slow to break down (accumulating in sediments of streams, rivers, lakes and coastal regions) and therefore still pose a serious health risk to anyone eating contaminated fish. See EDF Seafood Selector, Environmental Defense Fund, www.seafood.edf.org.
6 M. J. Zeilmaker et al., 'Fish consumption during child bearing age: A quantitative risk-benefit analysis on neurodevelopment', *Food and Chemical Toxicology*, 54 (2013), pp. 30–34.
7 Intensively reared broiler chickens are grown from hatchlings to maturity in six to nine weeks. This is like a human child putting on twice the weight of an adult (136 kg or 300 lb) in ten years. See Compassion in World Farming, 'The life of broiler chickens' (1 May 2013), www.cwf.org.uk.
8 Y. Wang et al., 'Modern organic and broiler chickens sold for human consumption provide more energy from fat than protein', *Public Health Nutrition*, 13, no. 3 (2010), pp. 400–8.
9 Researchers at Oxford University found a positive association between poultry intake and prostate cancer and Non-Hodgkin lymphoma: A. Knuppel et al., 'Meat intake and cancer risk: Prospective analyses in UK biobank', *Journal of Epidemiology and Community Health* (5 September 2019).
10 Andrew Purvis, 'It's supposed to be lean cuisine. So why is this chicken fatter than it looks?' *Observer* (15 May 2005).
11 The recommended level is 250 mcg per day from algae oil free from toxic contaminants. See Dr M. Greger, 'Should we take EPA & DHA omega 3 for our heart?' (16 August 2019), www.nutritionfacts.org..
12 A lecture by Dr Tim Radak, 'Do vegans and vegetarians require a DHA supplement?' (at the 45th Annual Conference of the North American Vegetarian Society, 6 July 2019, University of Pittsburgh at Johnstown, Pennsylvania) explains just how inconclusive studies are when it comes to the widely accepted benefits of eating fish and taking fish oil supplements.
13 See n. 4 for Dr Michael Greger's recommended intake of vitamin B12.
14 In a report by Grand View Research (2019) the global beef market alone is expected to be worth $383.5 billion by 2025: 'Beef market worth $383.5 billion by 2025' (January 2019), www.grandviewresearch.com.
15 'America's secret animal drug problem: How lack of transparency is endangering human health and animal welfare', Center for Food Safety, Executive Summary (September 2015), www.centerforfoodsafety.org.
16 NutritionFacts is a non-commercial, science-based public service providing free updates on the latest in nutritional research: www.nutritionfacts.org.
17 Dr M. Greger, 'Heart disease starts in childhood', NutritionFacts.org video, vol. 14 (23 September 2014).
18 K. Michaëlsson et al., 'Milk intake and risk of mortality and fractures in women and men: Cohort studies', *British Medical Journal* (28 October 2014).
19 Calcium in dark green leafy vegetables is absorbed twice as well as the calcium in milk and without all the baggage of high cholesterol and saturated fat. Leafy vegetables also come with fibre, folate, iron, antioxidants and bone-health superstar vitamin K, none of which are found in milk. See 'Plant vs. cow calcium' (5 September 2008) in www.nutritionfacts.org.
20 Dr M. Greger, 'Cow's milk casomorphin and autism', NutritionFacts.org video, vol. 8 (9 May 2012).
21 USDA, 'Milk quality, milking procedures, and mastitis on U.S. dairies' (2014), www.aphis.usda.gov.
22 Dr M. Greger, 'How much pus is there in milk' (8 September 2011), www.nutritionfacts.org.
23 When talking to children and young adults about this during outreach events, the look of disgust on their faces when they suddenly realize what they've been duped into is priceless. When talking to pre-millennials, the response is either

one of denial ('we've always done it so what's the problem?') or a very awkward silence that speaks volumes.
24 Artificial insemination (AI) involves a human pushing their arm into the anus of the cow to locate the cervix before inserting a device into the cow's vagina from which to inject bull semen.
25 Non-consensual AI, offspring being taken at birth, male calves shot and the eventual slaughter of all animals, free-range or otherwise. The film *Dismantle Dairy* (backed up by eighteen months of undercover filming at dairy farms throughout the UK) documents the systematic abuse that cows endure from humans taking their milk: www.dismantledairy.org.
26 'Dairy farm closures: More than 1000 shut in three years' (12 July 2016), www.bbc.co.uk; Agriculture and Horticulture Development Board, 'Producer numbers in England and Wales continue to tumble' (10 January 2019), www.dairy.ahdb.org.uk.
27 Sophie Gallagher, 'Has milk been put out to pasture? Young people (and Canada) are ditching dairy', *Huffington Post* (24 January 2019), www.huffingtonpost.co.uk.
28 Peter Wells, 'US's biggest milk producer files for bankruptcy protection', *Financial Times* (12 November 2019).
29 Based on more than 800 studies by the International Agency for Research on Cancer (26 October 2015). This does not mean that they are all equally dangerous. The classifications describe the strength of the scientific evidence about an agent being a cause of cancer rather than assessing the level of risk. An analysis of data from ten studies estimated that every 50 g portion of processed meat eaten daily increases the risk of colorectal cancer by about 18 per cent; see 'Q&A on the carcinogenicity of the consumption of red meat and processed meat' (October 2015), www.who.int.
30 Plant-based physician Dr Milton Mills explains that a carnivore or omnivore's digestive tract is much shorter so that rotting flesh can be digested quickly and excreted before it causes sickness. See Jennie Richards, 'Are humans designed to eat meat?' (20 July 2019), www.humanedecisions.com.
31 Reactions from the public when graphic footage is shown at Anonymous for the Voiceless events (see p. 176) range from disgust to aggression against activists showing it.
32 K. Smith, '"What The Health" Doctor Milton Mills says humans are vegan and aren't designed to eat meat' (September 2018), www.livekindly.co.
33 'National diabetes statistics report', Centers for Disease Control and Prevention (2017), www.cdc.gov.
34 N. D. Barnard et al., 'A low-fat vegan diet and a conventional diabetes diet in the treatment of type 2 diabetes: A randomized, controlled, 74-wk clinical trial', *American Journal of Clinical Nutrition*, 89 (2009), pp. 1588S–1596S.
35 T. Kuzuya, 'Prevalence of Diabetes Mellitus in Japan compiled from literature', *Diabetes Research and Clinical Practice*, 24 supplement (1994), pp. S15–S21.
36 S. Tonstad et al., 'Type of vegetarian diet, body weight and prevalence of type 2 diabetes', *Diabetes Care*, 32, no. 5 (2009), pp. 791–6.
37 Michelle McMacken and Sapana Shah, 'A plant based diet for the prevention and treatment of type 2 diabetes', *Journal of Geriatric Cardiology*, 14, no. 5 (2017), pp. 342–54. The Adventist Health Study 2 found that even small increases in red meat and poultry consumption in a cohort of relatively health-conscious subjects consuming meat no more than twice a week disproportionately increased the risk of type 2 diabetes.
38 I did try to keep in mind Douglas Adams's famous quip: 'I love deadlines. I love the sound they make as they go wooshing by.'
39 L. Baertlein and T. Polansek, 'Antibiotics sales for use in U.S. farm animals rose in 2015: FDA', Reuters (22 December 2016), www.reuters.com.
40 'Why non-therapeutic use of antibiotics in farm animals should end', Executive Summary, Compassion in World Farming (2011), ciwf.org.uk
41 'Antimicrobial resistance: Tackling a crisis for the health and wealth of nations; The review on antimicrobial resistance chaired by Jim O'Neill' (2014), www.amr-review.org.
42 It's estimated that the world currently produces enough food to feed ten billion people. See E. Holt-Giménez, 'We already grow enough food for 10 billion people . . . and still can't end hunger', *Journal of Sustainable Agriculture* (2012).
43 Erica Hellerstein and Ken Fine, 'A million tons of feces and an unbearable stench', *Guardian* (20 September 2017).
44 J. McWilliams, 'PTSD in the slaughterhouse', *Texas Observer* (7 February 2012), www.texasobserver.org.
45 A. J. Fitzgerald, L. Kalof and T. Dietz, 'Slaughterhouses and increased crime rates: An empirical analysis of the spillover from "The Jungle" into the surrounding community', *Organization and Environment*, 22, no. 2 (2009).
46 Chas Newkey-Burden, 'There's a Christmas crisis going on: No one wants to kill your dinner', *Guardian* (19 November 2018).
47 According to Visiongain, a market research company, the market for meat alternatives is set to reach $5.8 billion by 2022. See 'Meat substitutes market report 2018–2028' (21 March 2018), www.visiongain.com.
48 A collaborative application (with Heywood & Condie and Darryl Moore) to use this as a theme for a show garden at the Chelsea Flower Show is still on the backburner. If realized, it will present a unique insight into the future of gardening

with a hybrid environment fusing nature and technology, highlighting the power of plants and our relationship to the environment.

6. FOOD

1 Martha Busby, 'Greggs chief converted to veganism convinced of health benefits', *Guardian* (16 November 2019).
2 With Greggs's share prices rising 58 per cent in the six months to June 2019, it's clear that the momentum behind veganism is going to be much influenced by economics.
3 S. Bergman, 'The dairy indusrty to take legal action against vegan cheesemonger', *Independent* (12 February 2019).

7. TRANSITION

1 Thousands of dogs and cats are tortured and killed at this notorious festival, a fraction of the 10 million that are reported to be consumed each year in China. No one knows the exact numbers, but 'at its height, the Yulin Lychee and Dog Meat Festival was said to be responsible for 10,000 to 15,000 slaughtered dogs. In 2014 that figure was reported as dropping to 2,000 to 3,000. Reliable sources in 2016 – when the event was no longer official – put it at under 1,000', 'The truth about the Yulin Dog Festival – And how to stop it' (1 June 2018), www.animalsasia.com. See also M. Willett, 'This Chinese dog-eating festival's days are numbered thanks to a massive social media campaign' (30 June 2015), www.businessinsider.com.
2 H. Harwatt and M. N. Hayek, 'Eating away at climate change with negative emissions – Repurposing UK agricultural land to meet climate goals' (2019), https://animal.law.harvard.edu. Currently, 91 per cent of all cropland is used to grow only seven crops, most of which are used for animal feed.
3 Officially, the UK emits 1 per cent of GGE, but this is misleading as it only takes into consideration UK production emissions. It doesn't take into account the fact that we've outsourced much of our manufacturing to other countries (such as China). Arguably, a percentage of GGE embedded in our imports should be on our balance sheet.
4 Population in England increased from 5.7 million in 1750 to 16.6 million in 1850. See Mark Overton, *Agricultural Revolution in England: The Transformation of the Agrarian Economy, 1500–1850* (Cambridge University Press, 1996), summarized in British History (17 February 2011), www.bbc.co.uk.
5 Fallow land was typically ploughed to remove weeds, but growing turnips in rows allowed hand-hoeing while the crop was growing.
6 Bacteria attached to the roots of legumes transform atmospheric nitrogen into nitrates (seen as nodules) that are then exploited by the plants growing there in successive years.
7 'By 1850 only 22 per cent of the British workforce was in agriculture, the smallest proportion for any country in the world' (Overton, *Agricultural Revolution in England).*
8 H. R. de Ruiter et al., 'Total global agricultural land footprint associated with UK food supply 1986–2011', *Global Environmental Change*, 43 (March 2017), pp.. 72– 81. .
9 In January 2019 Tesco announced that it was closing fish, meat and deli counters in ninety of its stores due to lack of demand.
10 The said ram kept me sitting in a small oak for nearly an hour before he got bored and wandered off.
11 Forested land comprises 10 per cent in England, 15 per cent in Wales, 18 per cent in Scotland and 8 per cent in Northern Ireland. The European average is 38 per cent. See 'Forestry in England: Seeing the wood for the trees' (21 March 2017), www.publications.parliament.uk.
12 Ibid. See also O. Rudgard, 'England plants so few trees that the entire year's planting could have been done by three people', *Telegraph* (15 May 2017).
13 Based on the assumption of planting 1,000–2,000 trees per hectare.
14 'The thorn is the mother of the oak' (ancient saying).
15 Scrub provided medicines, dyes, food, fodder, thatching, furniture and gunpowder. See I. Tree, 'We need to bring back the wildwoods of Britain to fight climate change', *Guardian* (26 November 2018).
16 A. G. Tansley, *The British Islands and their Vegetation*, vols 1 and 2, 3rd edn (Cambridge University Press, 1953). See also F.W.M. Vera, *Grazing Ecology and Forest History* (CABI Publishing, 2000).
17 S. Elbein, 'Tree planting programs can do more harm than good', *National Geographic* (26 April 2019).
18 Ibid. Until trees are 60–80 years old (with their canopies out of reach of flames), they will 'burn through a 12 × 12 foot plantation like matchsticks'.
19 Ibid.
20 M. Rahman, 'Reclaiming the Rajasthan Desert from a voracious Mexican plant', *Guardian* (28 July 2014). .
21 Research by Silvoarable Agroforestry for Europe (SAFE) suggests that farm profitability can be increased by 10–50 per cent with high value trees such as walnut.
22 The procrastination is possibly due to concerns about post-Brexit agricultural policies yet to be thrashed out.
23 J. Tickell, *Kiss the Ground: How the Food You Eat Can Reverse Climate Change, Heal Your Body & Ultimately Save Our World* (Enliven Books, 2017).
24 George Monbiot (ed.), 'Land for the many: Changing the way our fundamental asset is used, owned and governed' (2019), https://landforthemanyuk.

25 What we now know as the suburbs were once thriving market gardens. Following the arrival of rail travel in the nineteenth century, market gardens prospered, delivering fresh fruit and vegetables quickly into town, but such land soon became attractive to housing development.
26 M. Vyawahare, 'World's largest vertical farm grows without soil, sunlight or water in Newark', *Guardian* (14 August 2016).
27 Up to thirty crop cycles are possible compared to three cycles on a conventional farm. With one of the densest populations in the world, Singapore produces only 7 per cent of the food it consumes.
28 J. Mateo-Sagasta, S. Marjani Zadeh and H. Turral, 'Water pollution from agriculture: A global review' (Food and Agriculture Organization of the United Nations, Rome, 2017, and the International Water Management Institute on behalf of the Water Land and Ecosystems Research Program, Colombo, 2017).
29 www.sadhanaforest.org. See also Auroville Green Practices, www.green.aurovilleportal.org.
30 Sadhana Forest achieved third place in the Humanitarian Water and Food Awards (Denmark, 2010).
31 A study by J. Poore and T. Nemecek, 'Reducing food's environmental impacts through producers and consumers', *Science* (1 June 2018), pp. 987–92, found that without meat and dairy consumption, global farmland use could be reduced by more than 75 per cent – an area equivalent to the USA, China, the European Union and Australia combined – and still feed the world.

8. ADVOCACY

1 See Melanie Joy, *Why We Love Dogs, Eat Pigs and Wear Cows* (2009).
2 Pigs in their natural habitat would normally spend 12–14 weeks with their mother.
3 The process of disbudding involves a hot iron being pressed into a calf's undeveloped horns to stop them from growing. In the UK it's a legal requirement to administer local anaesthetic before disbudding, but the RSPCA states that 'analgesia (pain relief) is sometimes not used for procedures such as dehorning mature animals'; see 'Beef cattle – Key welfare issues', www.rspca.org.uk. Also 3 per cent of beef is imported from Brazil where the dehorning and castration of cattle can be carried out without anaesthetic; see J. Jowit, 'Quarter of meat sold in UK imported from nations weaker on animal welfare', *Guardian* (16 February 2010).
4 'Sheep in the UK beaten, stamped on, cut, and killed for wool: A disturbing PETA Asia eyewitness investigation of the wool industry in the UK shows workers beating, stamping on, kicking, mutilating, and throwing around sheep': PETA [People for the Ethical Treatment of Animals] Asia, YouTube.
5 F. Lawrence, 'If consumers knew how farmed chickens were raised, they might never eat their meat again', *Guardian* (24 April 2016).
6 H. Saul, 'Hatched, discarded, gassed: What happens to male chicks in the UK', *Independent* (5 March 2015).
7 'Viva investigates: Conveyor belt of death', www.viva.org.uk.
8 A. Capps, 'The unavoidably violent history of backyard eggs' (18 February 2019), www.freefromharm.org.
9 According to the animal welfare charity PETA, the average breast of an eight-week-old broiler chicken is eight times heavier than it was twenty-five years ago.
10 Words by someone unknown on an Instagram post by @letfishlive.
11 The meat industry isn't exactly falling over itself to show consumers exactly what 'humane slaughter' looks like. Maybe some 'humane slaughter' footage could be shown at the meat counter so that customers can decide for themselves?
12 'The slaughter of farmed animals in the UK', www.viva.org.uk.
13 Ibid.
14 See *Land of Hope and Glory*, www.landofhopeandglory.com.
15 The joke made by William Sitwell (editor of *Waitrose Food* magazine) about vegans being force-fed meat, in response to a pitch by a vegan journalist, was more spiteful than humorous and sought to trivialize vegan issues and the associated violence. The ill-judged remark eventually cost him his job.
16 Dr Richard Oppenlander, 'The world hunger–food choice connection: A summary' (blog, 22 April 2012), www.comfortablyunaware.com.
17 On rare occasions, either through persistent dialogue or a vehicular accident where animals have fallen from lorries or trailers, the slaughterhouse owner might give up an injured or sickly animal as a gesture of goodwill, but this is an exception.
18 Peter Ak, @letallbejust, Instagram (2 January 2019).

9. A CALL TO GARDENERS

1 John Brookes, *A Landscape Legacy*, published by Pimpernel Press (2018).
2 Rosling's statistics show that the proportion of the world's population living in extreme poverty (level one, earning just $1 a day) has almost halved to one billion; the sobering reality is that five billion people (in levels two and three) live on an income of $4–16 a day.
3 Thanks to conservation initiatives, black rhino numbers have recovered from the brink of

extinction twenty years ago and number over 5,000 today. However, ongoing poaching for their horn means they are still considered critically endangered: 'a lot of work remains to bring the numbers up to even a fraction of what they once were – and to ensure that they stay there' (World Wildlife Fund, 'Black rhino', www.worldwildlife.org). While tiger numbers have recovered slightly (currently numbering approximately 3,890), they face unrelenting pressures from human activity (poaching, retaliatory killings and habitat loss) and are still on the WWF's endangered list. Giant pandas are still considered vulnerable by the WWF and could face even more challenges with China's bamboo forests threatened by climate change.

4 Currently only 15 per cent of the earth's surface and 3 per cent of the sea have conservation status. See Simon Worrall, 'Saving half the planet for nature isn't as crazy as it seems', *National Geographic* (27 March 2016), www.nationalgeographic.com.

5 Truth or Drought, 'Myth: Vegan diets waste marginal land', www.truthordrought.com.

6 Quoted in Worrall, 'Saving half the planet for nature isn't as mad as it sounds'.

7 Melanie Joy, *Why We Love Dogs, Eat Pigs and Wear Cows (2011).*

8 Joy has since updated this to four Ns to include 'nice'. See E. M. Barwick, 'The N-words meat eaters use' (18 May 2016), www.bitesizevegan.org.

9 Clare Mann, *Vystopia: The Anguish of Being Vegan in a Non-Vegan World* (Communicate 31 Pty Ltd, 2018).

10 Vegan Society, www.vegansociety.com.

11 Some have described humans as parasites, but even that falls short of the mark because parasites rely on keeping their host alive whereas we are actually killing the host that sustains us.

12 FAO, 'How to feed the world in 2050' (FAO Meeting of Experts, Rome, October 2009), www.fao.org.

13 B. Kelly, 'What is Earth Overshoot Day and why is it coming earlier each year?', *Independent* (31 July 2018).

14 'Climate change and land – An IPCC special report on climate change, desertification, land degradation, sustainable land management, food security and greenhouse gas fluxes in terrestrial ecosystems', www.ipcc.ch/report/srccl.

15 Prof. Walter Willett, MD, et al., 'Food in the Anthropocene: The EAT–Lancet Commission on healthy diets from sustainable food systems', *Lancet* (16 January 2019), It's interesting to note thst the World Health Organization withdrew its initial support of the EAT–Lancet report as a result of political pressure; see P. Ibirogba, 'World Health Organization withdraws support of plant-rich diet following political pressure' (13 April 2019) www.vegannews.co.

16 Neither of my letters received a response from the director.

17 'As the rainforest species disappear, so do many possible cures for life-threatening diseases. Currently, 121 prescription drugs sold worldwide come from plant-derived sources. While 25% of Western pharmaceuticals are derived from rainforest ingredients, less that 1% of these tropical trees and plants have been tested by scientists': Save the Amazon Coalition, 'The Disappearing Rainforests', www.savetheamazon.org. See also John Vidal, 'Protect nature for world economic security, warns UN biodiversity chief', *Guardian* (16 August 2010).

18 A little searching on the Internet is all it takes to learn the truth about how elephants are treated.

19 As a consequence of ingesting something alien from the river, I returned home in such poor health that I was off work for five weeks and hospitalized for five days. I look at it now as karmic retribution and on balance reckon I got off lightly compared with what the elephants have to suffer for our pleasure.

20 One of the first things I learned as a vegan is that personal choice isn't a personal choice when a victim is involved.

21 Renee King-Sonnen of Rowdy Girl Sanctuary; see Instagram: @rowdygirlsanctuary.

22 For instance, grass-fed meat has omega-3 DHA, something that needs thought to get into a vegan diet without taking a supplement.

23 Many people think that the ponies on Exmoor and Dartmoor live an idyllic life roaming free on vast tracks of land. The truth is that they are sold cheaply (some for as little as £10) and sent as live exports to Europe to be slaughtered for the leather industry.

24 'How many vegans are there in Great Britain?', www.vegansociety.com.

25 Also, in light of the health issues and associated costs to taxpayers associated with the consumption of animal products, it's not unreasonable to think that governments should impose VAT on meat and dairy products to help pay for the illness they cause.

FURTHER READING

Adams, Carol, J., *The Sexual Politics of Meat*, Continuum, 2010
Ahuja, Anjana, 'Are These Chickens of the Future', *Financial Times*, 19 February 2016
Appleby, Matthew, *The Super Organic Gardener: Everything You Need To Know about a Vegan Garden*, White Owl, 2018
Backster, Cleve, *Primary Perception: Biocommunication with Plants, Living Foods and Human Cells*, White Rose Press, 2003
Brown, Gabe, *Dirt to Soil: One Family's Journey into Regenerative Agriculture*, Chelsea Green, 2018
Francione, Gary, and Anna Charlton, *Eat Like You Care: An Examination of the Morality of Eating Animals*, Exempla Press, 2013
Greger, Michael, *How Not To Die*, Macmillan, 2016
Hall, Jenny, and Iain Tolhurst, *Growing Green: Organic Techniques for a Sustainable Future*, Vegan Organic Network, 2006, rev. edn 2010
Joy, Melanie, *Beyond Beliefs*, Roundtree Press, 2017
––, *Why We Love Dogs, Eat Pigs and Wear Cows*, Red Wheel Weiser, 2011
Leenaert, Tobias, *How to Create a Vegan World: A Pragmatic Approach*, Lantern Books, 2017
Mann, Clare, *Vystopia: The Anguish of Being Vegan in a Non-Vegan World*, Communicate 31 Pty Ltd, 2018
Monbiot, George, *Feral: Rewilding the Land, Sea and Human Life*, Penguin, 2014
––, *How Did We Get Into This Mess?*, Verso, 2016
Monbiot, George, et al., 'Land for the Many: Changing the Way Our Fundamental Asset is Used, Owned and Governed', 2019, https://landforthemany.uk
O'Connor, Carmen, *The Integrity of Love*, 2019, www.theintegrityoflove.com
Oppenlander, Richard, *Comfortably Unaware*, Beaufort Books, 2012
Postel, Sandra, 'Love Water for Chocolate', *National Geographic*, 12 February 2015
Pritikin, Robert, and Patrick McGrady, *The Pritikin Program for Diet and Exercise*, Bantam, 1979
Rackham, Oliver, *The History of the Countryside*, Phoenix, 1986
––, *Woodlands*, William Collins, 2015
Reynold, Mary, *The Garden Awakening: Designs to Nurture our Land and Ourselves*, Green Books, 2016
Singer, Peter, *Animal Liberation*, HarperCollins, 2009
Stallwood, Kim, *Growl: Life Lessons, Hard Truths, and Bold Strategies from an Animal Advocate*, Lantern Books, 2014
Stuart-Smith, Sue, *The Well Gardened Mind*, Collins, 2020
Tallamy, Douglas W., *Bringing Nature Home: How Native Plants Sustain Wildlife in Our Gardens*, Timber Press, 2007
Tickell, Josh, *Kiss the Ground, Enliven Books*, 2017
Tolhurst, Iain, *Back to Earth*, Tolhurst Organic Partnership, 2016
Tree, Isabella, *Wilding*, Picador, 2018
Vogt, Benjamin, *A New Garden Ethic: Cultivating Defiant Compassion for an Uncertain Future*, New Society Publishers, 2017
West, Cleve, *Our Plot*, Frances Lincoln, 2011
Wilson, Edward O., *Half Earth: Our Planet's Fight for Life*, Liveright Publishing, 2016

RESOURCES

If you need help transitioning to a vegan lifestyle, start with www.challenge22.com (it takes 21 days to kick a habit, hence 22) where you can get free support, including a mentor and advice from a clinical dietician if necessary.
There are some great recipes on the following websites:
Bosh!TV: www.bosh.tv
Lazy Cat Kitchen: www.lazycatkitchen.com
Tibits: www.tibits.co.uk
Vegan Society: www.vegansociety.com
Veganuary: www.veganuary.com

NUTRITION
A reliable (not for profit) resource for nutrition is Dr Michael Greger's NutritionFacts. Check out his website and various lectures on www.nutritionfacts.org.
Whole Food Plant-Based (WFPB): This is an independent, non-partisan, non-profit organization that empowers sustainable health for humans and the planet through a plant-based lifestyle. It connects human health (prevention and reversal of chronic disease) and planetary health (prevention and reversal of planetary destruction) as two co-dependent factors. It also supports and demonstrates that a plant-based structure can have multiple health, environmental and economic benefits towards a sustainably healthier humanity and planet – www.wfpb.org.

PHYSIOLOGY
YouTube video: *Meet Dr Milton Mills* by Mercy for Animals

VEGAN ETHICS AND ANIMAL EXPLOITATION
As far as vegan ethics is concerned, Earthling Ed's powerful speech is essential: https://www.youtube/Z3u7hXpOm58. Ideally this should be seen after watching his film *Land of Hope and Glory*, which shows standard practices in the UK farming industry: www.landofhopeandglory.org. This is free to watch online. It's graphic so you might want to watch it with a friend. There are two more graphic films showing the scope of animal abuse and exploitation: *Earthlings* (www.nationearth.com) and *Dominion* (www.dominionmovement.com). Another essential film to see is *Dismantle Dairy*, which shows the systematic exploitation of and cruelty to cows at dairy farms across the UK: www.dismantledairy.org.
Also, Gary Yourofsky's speech on YouTube has been pivotal in helping many people go vegan: 'Gary Yourofsky – The Most Important Speech You Will Ever Hear'. He also has a website: www.adaptt.org.
My advice would be to watch the following non-graphic films first as they should be reason enough to go vegan:
Cowspiracy examines the negative effects of animal agriculture on the environment: www.cowspiracy.com.
What the Health looks at how meat, dairy, fish and eggs cause some of the most common chronic diseases associated with the western world: www.whatthehealthfilm.com.
The Game Changers is a film by James Cameron which explores how top athletes are switching to plant-based diets to enhance their performance: www.gamechangersmovie.com.
Both *Cowspiracy* and *What the Health* are on Netflix. Despite the fact that all sources for the information included in these films are listed online, there have been claims that the director of both films, Kip Andersen, changed or altered facts to support the underlying vegan agenda. It's worth noting that those refuting the facts and figures in these films were invited to take part in a public online unedited debate. A few agreed to take part, but once Andersen tried to set up a date for the debate, they all pulled out. It turns out that studies and statistics used to refute the information in these films were funded by the animal agriculture industry. The fact that nutritionists and doctors get information from studies so obviously biased in favour of animal agriculture should be setting off alarm bells. Such misinformation is a deliberate marketing strategy to confuse consumers and protect profits.
As hard as they are to watch, I would still recommend seeing the graphic films so you have a complete understanding of the scale of exploitation and cruelty which the respective industries go to great lengths to conceal from their customers. *Land of Hope and Glory* is important as all of the footage is recent and from the UK. I would also advise seeing these films with a friend, partly as an act of self-preservation in terms of mental health and partly so that when you speak about issues highlighted in the films, you have support from someone who can verify that what you have seen is true (non-vegans can be quick to argue that such films are from isolated cases and from other countries only).
If you need any further advice or information, please don't hesitate to contact me by email: clevewest@btconnect.com.
Thank you and good luck!

OTHER USEFUL LINKS:

GENERAL

Plant Based News: www.plantbasednews.org

ADVOCACY

Animal Aid: www.animalaid.org.uk
Animal Rebellion: www.animalrebellion.org
Anonymous for the Voiceless:
www.anonymousforthevoiceless.org
DxE: www.directactioneverywhere.com
Hunt Saboteurs Association: www.huntsabs.org.uk
League Against Cruel Sports: www.league.org.uk
Let Fish Live: www.letfishlive.org
Mercy for Animals: www.mercyforanimals.org
Peta UK: https://www.peta.org.uk
The Save Movement: www.thesavemovement.org
Sea Shepherd: www.seashepherd.org.uk
Surge Activism: www.surgeactivism.com
Those Who Love Peace: www.twlp.org
Vegan Sidekick: Vegansidekick.com
Viva: https://www.viva.org.uk

VEGANICS

Aske Organic Farm: www.facebook.com/veganorganicfarm
Beans and Herbs: www.beansandherbs.co.uk
Brook End: www.emptycagesdesign.org
Centre for Alternative Technology (CAT):
www.cat.org.uk
Chyan Community Field: www.chyan.org
Debdale Eco Centre: debdale-ecocentre.org.uk
Drimlabarra Herb Farm: www.veganherbal.com
Growing with Grace: www.growingwithgrace.org.uk
Hulme Garden Centre:
www.hulmegardencentre.org.uk
No-dig gardening: www.charlesdowding.co.uk
PlantGrow – A natural, sustainable and chemical-free fertiliser produced by anaerobic digestion while creating electricity during the process.:
www.plantgrow.co.uk
Plants for a Future: www.pfaf.org
Scilly Organics: www.scillyorganics.com
Shumei Natural Agriculture Yatesbury
Yatesbury Natural Agriculture Farm:
www.shumei.eu/yatesbury
Tolhurst Organic: www.tolhurstorganic.co.uk
Vegan Organic Network: www.veganorganic.net
Welheath Community Co-op:
welhealthcommunitycooperative.wordpress.com

FOOD

The Fields Beneath: https://www.thefieldsbeneath.com
Little Ginger: http://www.littleginger.org/street-food
Love Seitan: https://www.loveseitan.com
Utta Nutta: http://www.uttanutta.com

COOKERY BOOKS

Carlin, Aine, *Keep it Vegan*, Kyle Books, 2014
––, *The New Vegan*, Kyle Books, 2015
Devi, Yamuna, *The Art of Indian Vegetarian Cooking*, Leopard Books, 1995
Firth, Henry, and Ian Theasby, *Bosh!*, HQ, 2018
Frei, Retro, Rolf Hiltl, and Juliette Chretien, *Vegan Love Story*, New Internationalist, 2015
Hingle, Richa, *Vegan Richa's Indian Kitchen*, Vegan Heritage Press, 2017
Jury, Jean-Christian, *Vegan*, Phaidon, 2017
Razavi, Parvin, *Vegan Recipes from the Middle East*, Grub Street, 2017
Sahni, Julie, *Classic Indian Vegetarian & Grain Cooking*, William Morrow, 1985
Sodha, Meera, *Fresh India*, Penguin, 2016

NUTRITION

Dr Esselstyn's Prevent and Reverse Heart Disease Program: www.dresselstyn.com
NutritionFacts: www.nutritionfacts.org
Physicians Committee for Responsible Medicine:
www.pcrm.org
Dr Michael Klapper MD (Physician, Speaker, Educator): www.doctorklaper.com

NUTRITION FOR COMPANION ANIMALS

The pet food industry, which is a big part of the meat industry, encourages us to feed our companion animals with food that they wouldn't naturally eat. Dogs are omnivores and can thrive on a plant-based diet. Cats are carnivores but some can make the transition to a plant-based diet and be healthier for it, as this report suggests: http://www.ncbi.nlm.nih.gov/pmc/articles/PMC5035952. You can get plant-based food for cats and dogs at www.veggiepets.com

ANIMAL SANCTUARIES

Bradley Nook Farm: https://www.facebook.com/BradleyNookFarm
Claresfarm: @claresfarm on Instagram
Friend Farm Animal Sanctuary: www.friendfarmanimalsanctuary.org
Hillside Animal Sanctuary: http://www.hillside.org.uk
Hugletts Wood Farm Animal Sanctuary:
@Huglettswoodfarm on Facebook
The Retreat Animal Rescue: www.retreatanimalrescue.org.uk
Tower Hill Stables Animal Sanctuary: http://www.towerhillstables.com

ANIMAL-FREE RESEARCH

Occasionally we are asked to sponsor someone doing great things to raise money for various charities. Since going vegan we have become aware that many of these charities support research where animals have been tested on. Clearly a vegan ethic can't support animal testing, so if you feel conflicted in this way you can donate instead to Animal Free Research UK where no animals are harmed: www.animalfreeresearchuk.org.

Some useful information about whether animal testing helps human medicine:

- Fewer than 2 per cent of human illnesses (1.16 per cent) are ever seen in animals. Over 98 per cent never affect animals.
- According to the former scientific executive of Huntingdon Life Sciences, animal tests and human results agree just '5–25 per cent of the time'.
- Among the hundreds of techniques available instead of animal experiments, cell culture toxicology methods give accuracy rates of 80–85 per cent.
- 92 per cent of drugs passed by animal tests immediately fail when first tried on humans because they are useless or dangerous.
- A 2004 survey of doctors in the UK showed that 83 per cent wanted an independent scientific evaluation of whether animal experiments had relevance to human patients. Fewer than 1 in 4 (21 per cent) had more confidence in animal tests than in non-animal methods.
- Rats are 37 per cent effective in identifying what causes cancer to humans – less use than guessing. The experimenters said: 'We would have been better off to have tossed a coin.'
- Rodents are the animals almost always used in cancer research. They never get carcinomas, the human form of cancer, which affects membranes (such as lung cancer). Their sarcomas affect bone and connective tissue; the two are completely different.
- Sex differences among lab animals can cause contradictory results. This does not correspond with humans.
- 75 per cent of side effects identified in animals never occur in humans.
- Over half of side effects suffered by humans cannot be detected in lab animals.
- Vioxx was shown to protect the hearts of mice, dogs, monkeys and other lab animals. It was linked to heart attacks and strokes in up to 139,000 humans.
- Genetically modified animals are not like humans. The mdx mouse is supposed to have muscular dystrophy, but the muscles regenerate with no treatment.
- The genetically modified CF mouse never gets fluid infections in the lungs: the cause of death for 95 per cent of human cystic fibrosis patients.
- In America, 106,000 deaths a year are attributed to reactions to medical drugs.
- In the UK an estimated 70,000 people are killed or severely disabled every year by unexpected reactions to drugs. All these drugs have passed animal tests.
- In the UK House of Lords questions have been asked regarding why unexpected reactions to drugs (which passed animal tests) kill more people than cancer.
- A German doctors' congress concluded that 6 per cent of fatal illnesses and 25 per cent of organic illness are caused by medicines. All have been animal tested.
- According to a thorough study, 88 per cent of stillbirths are caused by drugs which passed animal tests.
- 61 per cent of birth defects were found to have the same cause.
- 70 per cent of drugs which cause human birth defects are safe in pregnant monkeys.
- 78 per cent of foetus-damaging chemicals can be detected by one non-animal test.
- Thousands of safe products cause birth defects in lab animals – including water, several vitamins, vegetable oils, oxygen and drinking waters. Of more than 1,000 substances dangerous in lab animals, over 97 per cent are safe in humans.
- One of the most common lifesaving operations (for ectopic pregnancies) was delayed forty years by animal studies.
- The great Dr Hadwen noted, 'had animal experiments been relied upon . . . humanity would have been robbed of this great blessing of anaesthesia.'
- Aspirin fails animal tests, as do digitalis (heart drug), cancer drugs, insulin (which causes animal birth defects), penicillin and other safe medicines. They would be banned if animal testing results were believed.
- Blood transfusions were delayed two hundred years by animal studies.
- The polio vaccine was delayed forty years by tests on monkeys.
- 30 HIV vaccines, 33 spinal cord damage drugs and over 700 treatments for stroke have been developed in animals. None work in humans.
- Despite many Nobel prizes going to vivisectors, only 45 per cent agree that animal experiments are crucial.
- The Director of the Research Defence Society (which serves only to defend vivisection) was asked if medical progress could have been achieved without animal use. His written reply was, 'I am sure it could be.'

Source: www.vivisectioninformation.com.

INDEX

Page numbers in *italics* refer to illustrations.

ACKNOWLEDGEMENTS

First, I'd like to thank my publisher, Pimpernel Press, for their leap of faith in trusting in me to write this book; thanks in particular to Jo Christian, Gail Lynch, Emma O'Bryen and Becky Clarke. The fact that a couple of you started seriously considering veganism after reading the first few draft chapters was all the motivation I needed to meet my deadline. I'm also incredibly grateful to my editor, Nancy Marten, whose diligence and guidance has steered me through the potential pitfalls when dealing with such an emotive subject and has given the underlying message the best chance of being heard.

Second, thank you, the reader. It's not an easy book to read in terms of content, especially if you are a non-vegan. If you've read it all the way through with an open mind and without finding the notion of non-violence amusing, thank you again. I hope that it will inspire you to a more compassionate way of life that can have so many positive knock-on effects.

I'd like to say a big thank you to all the vegan advocates around the world who spend an incredible amount of their free time and hard-earned cash (yes, contrary to public perception, most of us do have jobs!) trying to educate people about vegan issues and paving the way to a kinder world. I'm especially grateful for the love and support from those that I have got to know at vigils and outreach events over the last few years. Meeting these people has restored some of my faith in humanity. I can't name them all and they won't expect me to (ego isn't their reason for speaking out against injustice), but I'd like to mention VeganSidekick, who has, in a creative way, helped many people see how illogical and unreasonable their arguments are for eating animals, and who kindly agreed to let me use a couple of cartoons for this book.

I'm grateful to Olivia Chapple for trusting me to design the first Horatio's Garden, which has helped me understand more fully the power of gardens as healing spaces and the joy that the natural world can bring to those who use them.

Dr Mahesh Shah, thank you for your help and inspiration on wholefood plant-based matters, and thank you Dr Helen Harwatt for your important work and guidance on environmental issues.

Thank you, Chaz Oldham, for introducing me to Barney and for making a potentially uncomfortable experience more bearable.

To Ruth, my wonderful assistant, thank you for putting up with the rants and raves, huffs and puffs, and giving me the time to write this book. Thank you to all the people in the gardening world who have contacted me about their own vegan journey. Your stories continue to inspire.

Huge love and thanks to my wife, Christine, and her vegan family, Joanna, Otis and Finton; Stephanie, Seb, Florence and not forgetting Otto, who made us all sit up and listen when he decided to go vegan at the age of twelve. Your support and encouragement have been vital. Joanna, you have taught us so much and remained stoic, patient and (outwardly at least) calm when we had our heads in the sand. I'm just so sorry we didn't come up for air sooner. Much love and thanks to Pete Shipp for allowing us to take over the kitchen and veganize the many wonderful celebrations we've had with you at your home. To my brother Andre, Diana and the boys, your love and support over the last few years has been crucial and much appreciated.

A special thanks to Christine for the wonderful drawings of animals I have witnessed moments before slaughter. I underestimated just how painful it would be for you to spend so many hours looking through photos of the victims over the last few years. Your observations in capturing their characters and acknowledging them as someone, not something, are a lasting tribute to them. They will never be forgotten. Thank you too for your love and understanding when my trust in humanity sank to an all-time low, not to mention your encouragement and patience while putting this book together.

If you've got this far and still think that vegans are forcing their beliefs on you, please remember that personal choice doesn't come into it when there's a victim involved. Vegans, just like people who are against whaling, bullfighting, trophy hunting, fox hunting or the Yulin Dog Festival, aren't forcing their beliefs on anyone. They're simply asking you not force your beliefs on others who can't speak up for themselves and who, given the choice, would rather live.

If you have been affected by any of the issues raised in this book, I'm always happy to help anyone who wants to change to a vegan lifestyle. Feel free to contact me through any of my social media platforms.

Finally, if you're inspired to go vegan but nervous about what your friends, family and colleagues will think, remember that this isn't about you. This is about trillions of beings that have no say in how their lives should be lived. This is about speaking up for the vulnerable, the oppressed and the voiceless victims of the world who exist as numbers, products and commodities. By helping them you are also speaking up for the health of our planet and that of our families. All you have to do is put yourself in the victim's place and imagine what it would be like to be born with the date for your execution already set. Wouldn't you want someone to speak up for you?

'TAKE ACTION WITH COMPASSION AS THE ROOT INTENTION AND YOU WILL PROVIDE INCREDIBLENESS. YOUR EGO WILL FIGHT YOU AS IT BEGINS TO LOSE ITS GRIP, BUT STAY STRONG, YOU CAN FEED COMPASSION INSTEAD.'

CARMEN O'CONNOR,
The Integrity of Love

PICTURE CREDITS

The publishers have made every effort to contact holders of copyright works. Any copyright holders we have been unable to reach are invited to contact the publishers so that a full acknowledgement may be given in subsequent editions. For permission to reproduce the images on the pages listed below the publishers would like to thank the following.

Front cover and p88 © Triff/Shutterstock
Page 2 © guildfordanimalsave
Page 11 © Chaz Oldham
Page 12 © Mike Russell/Shutterstock
Page 36 © Mooauan/Shutterstock
Page 39 © M&G Investments
Page 60 © 2happy/Shutterstock
Page 87 © Anthony Coleman
Page 106 © Kathie Nichols/Shutterstock
Page 124 © Fly Dragon Fly/Shutterstock
Page 140 © Triff/Shutterstock
Page 144 © Jay Wilde
Page 160 © logoboom/Shutterstock
Page 184 © Juraj Kovac/Shutterstock